# Pushing Electrons

## A Guide for Students of Organic Chemistry

**Third Edition**

**Daniel P. Weeks, Ph.D.**
Northwestern University

Harcourt College Publishers
Fort Worth Philadelphia San Diego New York Orlando Austin
San Antonio Toronto Montreal London Sydney Tokyo

Vice President/Publisher: John Vondeling
Product Manager: Angus McDonald
Developmental Editor: Sandra G. Kiselica
Project Editor: Sally Kusch
Production Manager: Alicia Jackson
Cover designer: Chazz Bjanes

Web Site Address
http://www.harcourtcollege.com

Printed in the United States of America

Pushing Electrons, third edition

ISBN: 0-03-020693-6
Library of Congress Catalog Card Number: 97-68076

0 1 2 3 4 5 6 7 8 9 039 16 15 14 13 12 11 10 9 8 7

To my sons,
John and Timothy

# *Postcard from a Student*

Dear Dan,

Last night I was sitting on the shore of San Francisco Bay watching the fog roll in and suddenly I understood what organic chemistry was all about.

Ric

# Preface

Among the factors that set organic chemistry apart from the other traditional divisions of chemistry are the endless variety of structure, the strong dependence on the resonance theory, and the emphasis on mechanisms. As students become more facile with structure and more adroit at writing resonance structures and mechanisms, organic chemistry changes from a bewildering array of facts to a unified science. Newly encountered reactions can be understood from previously learned principles.

At the introduction of resonance or mechanisms, the professor begins using little arrows that resemble fishhooks (⌒, ⌒). Using these arrows, electrons are moved around molecules and appear—as if by wizardry—precisely where the professor needs them. The penchant for moving electrons has earned organic chemists the pejorative *electron pusher* that, like *Yankee Doodle,* we wear proudly.

I have observed that students avoid pushing electrons. They find it difficult, therefore, to write resonance structures and lose a powerful tool to explain reactivity. Many students never become comfortable with bond-making and bond-breaking steps in organic mechanisms. With this low-tech, pencil-and-paper program, a student can learn to push electrons with the same wizardry as the professor.

While working through this book, the student will learn to construct Lewis structures of organic molecules and functional groups. The functional groups will be learned early and in a way that minimizes the necessity for memorization. Then a systematic method is presented whereby all the possible resonance structures for a molecule, ion, or radical can be generated by beginning with one Lewis structure and moving electrons.

Chapter 3 demonstrates the individual bond making and bond breaking steps in reaction mechanisms. Chapters 4 and 5 are new to this edition. In Chapter 4 students newly acquired ability to push electrons is used to construct complete mechanisms. The book concludes with a chapter in which electron pushing skills are applied to the reactions of biochemistry. The programmed approach of this book emphasizes repetition and active participation in learning how to push electrons.

In the production of three editions of the book, I have received help from Dr. Joyce C. Brockwell and Professor Emeritus Charles D. Hurd (Northwestern University), Dr. Rebecca Dickstein (Drexel University), Dr. Kenneth Turnbull (Wright State University), and Dr. Darrell J. Woodman (University of Washington). While retirement to rural western Virginia has its advantages, I needed access to a chemistry library and vigorous chemistry faculty in order to write the third edition. I found both at Virginia Polytechnic Institute and State University (Virginia Tech).

Professor Richard Gandour, Chemistry Department Head, provided a generous welcome. As always, I am grateful to John Vondeling and Sandi Kiselica of Saunders College Publishing for their encouragement and patience.

Daniel P. Weeks
June 1997

# To the Professor

Students will be better at organic chemistry if they master the skills in *Pushing Electrons.* The book is a supplement. It knows its place. It is short, methodical, and repetitious. A student will complete *Pushing Electrons* in 12 to 16 hours and have the rest of the time for important things such as listening to you and studying the main text.

The approach is methodical, the pace is slow. Strong students will race through certain sections, but even they will benefit. I tell them in the text, "you can never be too skilled at pushing electrons." Weaker students will appreciate that the author still holds a grudge against his calculus professors who skipped steps.

The repetition used in teaching students how to write Lewis structures and how properly to push electrons has an additional benefit. Students become comfortable writing and rewriting organic structures. The book may not eliminate pentavalent carbon but it should reduce the frequency of its appearance. The structures in this book, except for rings and larger structures in the last two chapters, show all carbons and hydrogens. I assert that the early introduction of line-angle structures does students great harm.

Assign Chapter 1, Lewis Structures, right away. Chapters 2 and 3 are relatively independent of one another; assign them as you see fit. It is not necessary for students to complete Chapter 2 before beginning Chapter 3. Although a chapter may be interrupted, it should be done from beginning to end.

There are two significant additions to the third edition. I have expanded the treatment of biochemistry in Chapter 5. Consider this as optional depending on whether or not your course contains biochemistry. Chapter 4 describes strategies for dealing with mechanism problems. I suggest that you assign parts of it as your students encounter mechanism problems of increasing difficulty. Once again, you may interrupt Chapter 4 but assign it from beginning to end.

# To the Student

I have three instructions. First, supply an answer wherever a line appears under a blank space. Actually, in Chapter 4 only blank spaces are provided, but by that time you will know what is required. The correct answer will be a word, a number, a structure, some arrows, or—in Chapter 4—a mechanism. Second, don't look up the correct answers until you have made a serious try at doing it yourself. Third, if you plan to just look up the right answers and transcribe them, return the book and get your money back.

This program uses two effective learning devices: active involvement and repetition. You will participate actively in the learning process. Because so much of the academic experience consists of receiving information, it should be refreshing to work through a program using your own wits. You will see an example of an operation and then carry it out several times as the supporting material is gradually removed. The approach is methodical. Some of you will find that you can accelerate your trip through certain sections. But the program has been written in the hope that none of you will ever feel abandoned. Expect to spend a total of 12 to 16 hours to complete the program.

I hope your professor will assign appropriate sections of the book as the course proceeds. Failing that, do Chapter 1 right away. Then, start Chapters 2 and 3 (they are relatively independent of each other). If you start to feel lost, put it aside for awhile and then pick up where you left off. Begin Chapter 4 when you need some help with solving mechanism problems. Chapter 5 is for courses where some biochemistry is introduced toward the end. From time to time during the course, I think you will find *Pushing Electrons* useful for review. This is especially true of Chapter 3, the backbone of the book.

# Contents

# Glossary

*Boldfaced numbers refer to the page on which the term first appears.*

**Aliphatic** One of two broad classes of organic compounds. The other is *aromatic*. Aliphatic compounds are composed of chains of carbon atoms. A subgroup within this class is the alicyclic compounds that contain rings of carbon atoms but are similar to open-chain compounds in chemical and physical properties. Actually, the most sensible definition of aliphatic compounds is "compounds that are not aromatic." **16**

**Aromatic** One of two broad classes of organic compounds. The other is *aliphatic*. Benzene, compounds containing a benzene ring, and compounds similar to benzene in chemical properties constitute the aromatic compounds. **16**

**Carbanion** An ion in which a negative charge resides on a carbon atom. **34**

**Carbocation** An ion in which a positive charge resides on a carbon atom. **34**

**Decarboxylation** A reaction in which a carboxylic acid loses carbon dioxide. **164**

**Double bond** The combination of a sigma bond and a pi bond between two atoms. The carbon–carbon double bond is written C=C. In this notation there is no convention regarding which line represents the sigma and which, the pi bond. The presence of a double bond makes a compound unsaturated. **7**

**Electron attracting and electron releasing groups** Various groups of atoms found in organic compounds are electron attracting or releasing, relative to the carbon atom to which they are attached. The presence of these groups can have a marked effect on chemical reactivity. Some electron releasing groups are: $-NH_2$, $-OCH_3$, and $-CH_3$. Some electron attracting groups are: $-NO_2$, $-COOH$, $-F$, $-Cl$, $-Br$, and $-CN$. **74**

**Electrophile** The word comes from the Greek meaning "electron lover." Electrophiles are chemical entities (atoms, ions, radicals, or molecules) that seek electrons. Electrophiles are electron-poor. They are designated as E or E+. When an electrophile reacts with some electron-rich center on an organic molecule, the process is called electrophilic attack. **72**

**Formal charge** A Lewis structure is, after all, only a graphic representation of a molecular structure. As useful as they are, Lewis structures are not entirely descriptive of a molecule. The formal charge is the charge that would be on an atom in a molecule if the Lewis structure were an entirely accurate representation of the molecule. In one sense, formal charge is simply electron bookkeeping. However, the location of formal charge on a Lewis structure is always revealing of the chemistry of the molecule. **20**

**Free radical** A chemical species having at least one unpaired valence electron. **36, 58, 102**

**Functional group** One can think of organic compounds as being composed of two parts: a hydrocarbon backbone and a functional group. The latter is an atom or group of atoms that confer some distinct chemical or physical property. The functional group is the site of chemical reactions. Thus, molecules with the same functional group undergo similar chemical reactions despite having quite different hydrocarbon backbones. **22**

**Grignard reagent** A reagent formed by reaction of an alkyl halide with magnesium in an ether solvent. These reagents are extremely useful and versatile intermediates for synthesizing organic compounds. The actual structure of the Grignard reagent is rather complex. It is expedient, however, to consider it an ion pair, as this text does. **97**

**Heterolytic cleavage (heterolysis)** The breaking of a bond so that the electrons that formed the bond are distributed unevenly between the two fragments. Heterolysis results in an anion and a cation. **27**

$$A{-}B \longrightarrow A^{+} + :B^{-}$$

or

$$A{-}B \longrightarrow A:^{-} + B^{+}$$

The counterpart of heterolysis is *homolysis,* in which the electrons are distributed evenly between the two fragments. The result is a pair of radicals.

$$A{-}B \longrightarrow A\cdot + \cdot B$$

**Metabolism** The sum total of all biochemical reactions that take place in an organism. **185**

**Nucleophile** The word comes from the Greek, meaning "nucleus lover." Nucleophiles are chemical entities (ions or molecules) that seek an electron-poor center. Nucleophiles are electron-rich. They are designated as Nu or $Nu^-$. When a nucleophile reacts with some electron-poor center of an organic molecule, the process is called *nucleophilic attack.* **72**

**Pi bond and pi electrons** Bonds that are the result of lateral (side to side) overlap of atomic p-orbitals. Pi bonds are weaker than sigma bonds. Pi bonds are always found in combination with a sigma bond. Therefore, single bonds are never pi bonds. Pi electrons are the electrons in pi bonds. **50**

**Carbocation** Carbocations are classified as primary, secondary, or tertiary according to the following scheme.

**Primary** The carbon atom with the positive charge is *directly* attached to only one other carbon atom and two hydrogen atoms. For example,

$$CH_3—\overset{+}{C}H_2$$

**Secondary** The carbon atom with the positive charge is directly attached to two other carbon atoms and one hydrogen atom. For example,

$$CH_3—\overset{+}{C}H—CH_3$$

**Tertiary** The carbon atom with the positive charge is directly attached to three other carbon atoms and no hydrogen atoms. For example,

$$CH_3—\underset{\substack{|\\ CH_3}}{\overset{\substack{CH_3\\ |}}{C^+}}$$

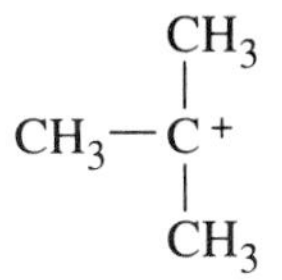

The order of stability of carbocations is tertiary > secondary > primary. **123, 124**

**Protonation** The combination of a proton (hydrogen ion) with some organic base to give the conjugate acid of that base. **90**

$$H^+ + :B \longrightarrow H—B^+$$

**Regiochemistry** The study of reactions in which one direction of bond making or bond breaking is preferred over all others. **166**

**Saturated** Compounds that contain only single bonds cannot add other reagents. Thus, they are called *saturated.* **1**

**Secondary carbocation** (See Carbocation.)

**Sigma bond and sigma electrons** Bonds that are the result of axial overlap of atomic orbitals. All single bonds are sigma bonds. Sigma electrons are the electrons in sigma bonds. **88, 96**

**$S_N1$ and $S_N2$ reactions** Organic reactions can be classified according to their mechanisms. The $S_N1$ and $S_N2$ reactions are probably the best known of these. This classification gives: the result, the type of reagent, and the molecularity of the reaction. Thus, $S_N1$ stands for "substitution, nucleophilic, unimolecular," and $S_N2$ stands for "substitution, nucleophilic, bimolecular." **88, 95**

**Stereochemistry** The study of the arrangement of atoms in three-dimensional space and how the arrangement affects the chemical and physical properties of molecules. **95, 170**

**Steric strain** Strain (instability) present in a molecule because of the arrangement of the atoms. Steric strain usually arises because bond angles are forced to depart from their ideal value, or because nonbonded atoms are required to be too close to one another. **49**

**Tertiary carbocation** (See Carbocation.)

**Unsaturated** Compounds that contain double or triple bonds are capable of adding other reagents. Thus, they are called *unsaturated.* **6**

**Unshared electrons** Valence electrons in a molecule that are not involved in binding one atom to another. In a Lewis structure they are the exclusive property of one atom. They are also called *nonbonding* electrons. **4**

# 1

# Lewis Structures

*In a perfect world there would be no need for this chapter at the beginning of a course in organic chemistry. The skills described here are taught in general chemistry. Skip this chapter if you feel entirely confident that you are able to write Lewis structures correctly and are comfortable with the concepts of ions, radicals, multiple bonding, and formal charge. On the other hand, these skills are vital to organic chemists and a little review never hurt anyone.*

The distinguished American chemist, Gilbert Newton Lewis (1875–1946), was the first to propose that atoms share a pair of electrons to form a covalent bond. Lewis electron-dot formulas are familiar to all students of introductory chemistry. Lewis used dots to denote all valence electrons in chemical formulas. We will use the current convention that denotes a pair of shared electrons (a covalent bond) as a dash connecting the atoms and shows only the unshared electrons as dots.

Using Lewis's notation together with a judicious application of the octet rule allows us to write useful chemical formulas. The most important requirement for the formation of a stable compound is that the atoms achieve a closed shell, noble gas configuration. For all but the very light elements, this means eight electrons. The light elements (H, Li, Be) require two electrons, the configuration of helium. A few compounds such as borane ($BH_3$), boron trifluoride ($BF_3$), and aluminum chloride ($AlCl_3$), which are important to elementary organic chemistry, do not conform to the octet rule.

## Molecules and Functional Groups

### Saturated Groups

Proper Lewis structures for **saturated** compounds can be written by following this brief set of rules.

#### RULE 1

---

*Write the molecular skeleton.* Historically, the correct skeleton of each functional group was deduced from experiment. Most of that has been done, however, and we shall assume that you know, or will know shortly, that methane's skeleton is

```
        H
    H   C   H        and not      H   H   C   H   H
        H
```

and formic acid's molecular skeleton is

```
        O
    H   C   O   H    and not      H   C   O   O   H
```

which is often written to save space. In your initial study of organic chemistry you must be careful to not let space-saving notation mislead you about structure.

## RULE 2

*Assume that all bonds are covalent.* This assumption is not always accurate, but it works most of the time, especially for organic compounds. As your experience increases, you will learn when to use ionic bonds.

## RULE 3

*Count the available valence electrons.* Each atom in a compound brings into the union a certain number of valence electrons. The number is determined by the atom's Group in the periodic table.

1. Hydrogen is a Group __I__ element and each hydrogen will contribute __1__ valence electron. Carbon is a Group ____ (Roman numeral) element and each carbon will contribute ____ (number) electrons. Every oxygen atom in a compound will contribute ____ valence electrons.

   Chloromethane has the molecular formula $CH_3Cl$. Its skeleton is

   ```
           H
       H   C   Cl
           H
   ```

   and the number of valence electrons may be determined as follows. There are three hydrogen atoms, each of which contributes __1__ valence electron; the single carbon contributes

__4__ electrons; and the single chlorine atom contributes __7__ electrons, making a total of __14__. A convenient tabular form of this calculation is

| | | | |
|---|---|---|---|
| 3 | H | $1 \times 3 =$ | 3 |
| 1 | C | $4 \times 1 =$ | 4 |
| 1 | Cl | $7 \times 1 =$ | 7 |
| | | | 14 |

2. Methanol has the molecular formula $CH_4O$. Its skeleton is

```
    H
H   C   O   H
    H
```

Each of four hydrogen atoms contributes ____ valence electron, the carbon atom contributes ____, and the oxygen atom contributes ____, making a total of ____.

| | | | | |
|---|---|---|---|---|
| ______ | H | ____________ | = | |
| ______ | C | ____________ | = | |
| ______ | O | ____________ | = | ______ |

## RULE 4

*Add electrons to the skeleton by making single bonds between the atoms that are bonded and then by providing each atom with a complement of eight electrons* (hydrogen requires only two electrons).

3. The skeleton of chloromethane is

________________

The central carbon atom is bonded to each of the other atoms by a shared electron pair (represented by a straight line, —) giving

```
    H
    |
H — C — Cl
    |
    H
```

Now, each hydrogen has two electrons and the carbon atom has eight. However, chlorine must be provided with **unshared electrons** (represented by pairs of dots, ••) to complete its octet, thus,

```
      H
      |    ••
  H — C — Cl:
      |    ••
      H
```

4. Methanol's skeleton is

____________________

Connecting all bonded atoms by means of an electron pair (single bond) gives

____________________

and completing the octet of oxygen with two pairs of dots gives

____________________

## RULE 5

*Count the electrons in the Lewis structure and compare the result with the number derived from Rule 3. If the two numbers are the same, the Lewis structure is correct.*

5. The structure for chloromethane is

```
      H
      |    ••
  H — C — Cl:
      |    ••
      H
```

It contains <u>14</u> valence electrons. The number of available valence electrons in chloromethane is _____. The Lewis structure is <u>correct.</u>

6. The structure for methanol is

```
        H
        |    ••
    H — C — O — H
        |    ••
        H
```

It contains _____ valence electrons. The number of available valence electrons is _____ . The structure is _______________ (correct, incorrect).

## Exercises

Using the method outlined above, derive the Lewis structures for the following compounds. The unbonded skeletons are provided.

7. Dimethyl ether ($C_2H_6O$)

**Rule 3**

______ H ____________ =

______ C ____________ =

______ O ____________ = ______

═════

**Rule 4**

```
    H   H
H   C O C   H
    H   H
```

**Rule 5**

No. of electrons in structure ______

No. of valence electrons ______

Structure is ____________(correct, incorrect)

8. Methylamine ($CH_5N$)

**Rule 3**

______ H ____________ =

______ C ____________ =

______ N ____________ = ______

═════

**Rule 4**

```
   H   H
H  C N
   H   H
```

**Rule 5**

No. of electrons in structure ______

No. of valence electrons ______

Structure is ___________

9. Methanethiol ($CH_4S$)

**Rule 3**

| | | | | |
|---|---|---|---|---|
| ______ | H | ___________ | = | |
| ______ | C | ___________ | = | |
| ______ | S | ___________ | = | ______ |
| | | | | ====== |

**Rule 4**

```
     H
H    C    S    H
     H
```

**Rule 5**

No. of electrons in structure ______

No. of valence electrons ______

Structure is ___________

10. Methylal ($C_3H_8O_2$)

**Rule 3**

| | | | | |
|---|---|---|---|---|
| ______ | H | ___________ | = | |
| ______ | C | ___________ | = | |
| ______ | O | ___________ | = | ______ |
| | | | | ====== |

**Rule 4**

```
    H       H       H
H   C   O   C   O   C   H
    H       H       H
```

**Rule 5**

No. of electrons in structure ______

No. of valence electrons ______

Structure is ___________

## Unsaturated Groups

11. When the compound in question is **unsaturated**, the application of Rule 5 shows that the number of valence electrons in the trial structure is larger than the number of available valence electrons. In such cases the trial structure cannot be correct. It must be modified to contain the proper number of electrons. The skeleton of ethylene ($C_2H_4$) is

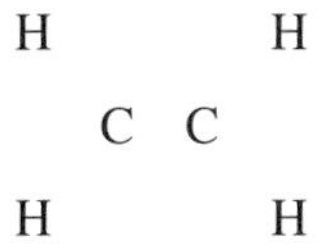

```
H       H
   C C
H       H
```

Each of 4 hydrogen atoms will contribute ____ electron and each of ____ carbon atoms will contribute ____ electrons. The total number of available valence electrons is ____.

$$\begin{array}{cccc} \underline{4} & H & \underline{1 \times 4} = & \underline{4} \\ \underline{2} & C & \underline{4 \times 2} = & \underline{8} \\ & & & \underline{\underline{12}} \end{array}$$

12. Adding electrons to the skeleton by making single bonds between all bonded atoms gives

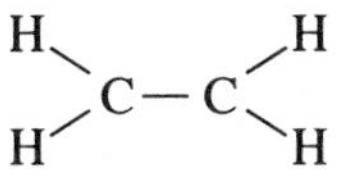

```
H       H
 \     /
  C — C
 /     \
H       H
```

Each hydrogen atom now has a pair of electrons, but each carbon has only 6 electrons. Adding a pair of electrons to each carbon gives the trial structure

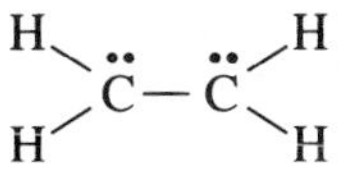

```
H  ..   ..  H
 \         /
  C  —  C
 /         \
H           H
```

The number of electrons in the trial structure is ____. Since this exceeds the number of available valence electrons the structure is incorrect.

## RULE 6

*When the number of electrons in the trial structure is larger than the number of available valence electrons, the structure may be corrected by introducing one or more multiple bonds.*

13. This is done by removing an unshared pair from each of two adjacent atoms and adding one electron pair as a second bond between the atoms. Each such operation reduces the number of electrons in the trial structure by two. Removing the unshared pairs of electrons on the carbon atoms and adding a second carbon–carbon bond gives

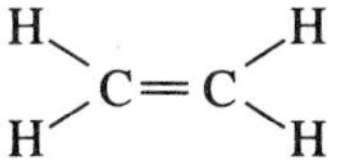

```
H       H
 \     /
  C = C
 /     \
H       H
```

a structure in which there are four electrons involved in a **double bond** between the carbon atoms. The trial structure now contains ____ electrons and is correct.

14. Formaldehyde has the skeleton

```
H
   C  O
H
```

Each of ____ hydrogen atoms will contribute ____ electron, the carbon atom, ____ electrons, and the oxygen, ____ electrons. The total number of valence electrons is ____.

| | | | | |
|---|---|---|---|---|
| ____ | H | ________ | = | |
| ____ | C | ________ | = | |
| ____ | O | ________ | = | ____ |
| | | | | ===== |

Adding single bonds to the skeleton gives

____________

and providing the carbon atom and the oxygen atom with an octet of electrons gives

____________

The number of electrons in the trial structure is ____. The structure is ____ (correct, incorrect). Since the trial structure is incorrect, the pair of unshared electrons on the carbon atom and one of the pairs of unshared electrons on the oxygen atom are removed. Adding a second carbon–oxygen bond gives

____________

The trial structure now contains ____ electrons and is ________ (correct, incorrect).

15. The skeleton of acetonitrile is

```
    H
H   C   C   N
    H
```

Each of ____ hydrogens will contribute ____ electron, each of ____ carbons will contribute ____ electrons, and the nitrogen will contribute ____ electrons. The total number of valence electrons is ____.

| | | | | |
|---|---|---|---|---|
| ____ | H | ________ | = | |
| ____ | C | ________ | = | |
| ____ | N | ________ | = | ____ |
| | | | | ==== |

Adding single bonds to the skeleton gives

________________

and providing the carbon and nitrogen atoms with octets gives

________________

The number of electrons in the trial structure is ____. The trial structure is ____ (correct, incorrect). Removing one unshared pair of electrons from carbon and one pair from nitrogen and adding a second carbon–nitrogen bond gives

________________

The structure now contains ____ electrons and is ____ (correct, incorrect).

Performing the same operation again gives

____________________

The trial structure now contains _____ electrons and is _____ (correct, incorrect).

Thus, you continue to remove pairs of unshared electrons from adjacent atoms and add multiple bonds (add unsaturation) until the number of electrons in the trial structure is equal to the number of available valence electrons.

*   *   *   *   *

Sometimes, you must decide where a multiple bond is best added.

16. Formic acid has the skeleton

```
      O
H C
      O H
```

The number of available valence electrons is _____ .

| | | | | |
|---|---|---|---|---|
| _____ | H | __________ | = | |
| _____ | C | __________ | = | |
| _____ | O | __________ | = | _____ |
| | | | | ===== |

Filling in the skeleton with single bonds gives

```
        O
       /
  H—C
       \
        O—H
```

and adding the appropriate unshared pairs gives

```
        :Ö:
   ..  /
  H—C
       \  ..
        O—H
        ..
```

The number of electrons in the structure is _____, which is __2__ too many electrons. This structure can be corrected by removing two unshared pairs and making one double bond. However, the double bond could be placed between the carbon atom and either one of the two oxygens. Thus,

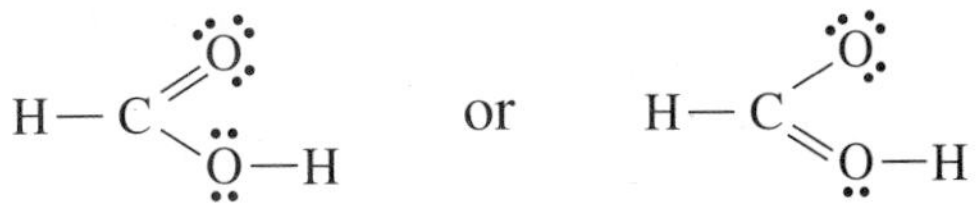

$$\mathrm{H{-}C}\overset{\ddot{O}:}{\underset{\ddot{\underset{..}{O}}-H}{\diagup\!\!\diagup}} \qquad \text{or} \qquad \mathrm{H{-}C}\overset{:\ddot{O}:}{\underset{\underset{..}{O}-H}{}}$$

The correct choice is the first structure and is made by observing that both oxygen atoms in the first structure have the appropriate valence of two. The second structure is less acceptable since it requires that the oxygen atoms have the unfamiliar valences of one and three respectively. We shall see later that the second structure is also less acceptable because it contains charge separation.

The elements most commonly found covalently bonded in organic compounds are listed below along with their valences.

| | |
|---|---|
| carbon and silicon | 4 |
| hydrogen | 1 |
| oxygen and sulfur | 2 |
| the halogens (F, Cl, Br, I) | 1 |
| nitrogen and phosphorous | 3 (4 in onium salts) |
| boron | 3 |

One determines the valence of an atom in a structure by counting the number of electron pairs it is *sharing* with other atoms. That an atom may be sharing more than one pair with another atom (double or triple bond) does not change this. Thus, carbon may exhibit a valence of four in the following ways.

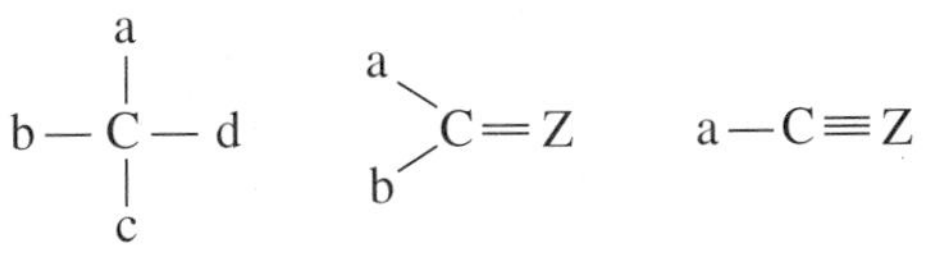

$$\mathrm{b{-}\underset{c}{\overset{a}{C}}{-}d} \qquad \overset{a}{\underset{b}{>}}\mathrm{C{=}Z} \qquad \mathrm{a{-}C{\equiv}Z}$$

Oxygen may exhibit its valence of two as follows.

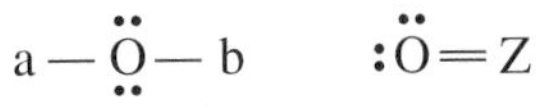

$$\mathrm{a{-}\ddot{\underset{..}{O}}{-}b} \qquad \mathrm{:\ddot{O}{=}Z}$$

Nitrogen may show a valence of three as follows.

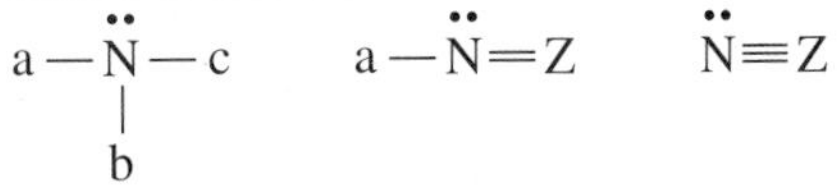

$$\mathrm{a{-}\underset{b}{\ddot{N}}{-}c} \qquad \mathrm{a{-}\ddot{N}{=}Z} \qquad \mathrm{\ddot{N}{\equiv}Z}$$

17. The skeleton of acetyl chloride is

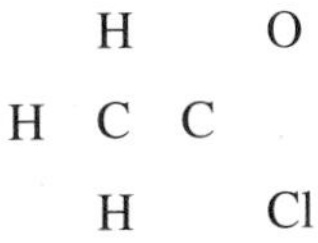

```
H     O
H  C  C
H     Cl
```

write the best Lewis structure for acetyl chloride by following the procedure in the previous problem.

## Exceptions to the Octet Rule

There are stable compounds that do not obey the octet rule. A few of them are important players in organic chemistry.

Aluminum trichloride and boron trifluoride have only six electrons in the outer shell of the central atom.

$$AlCl_3 \qquad\qquad BF_3$$

Although stable under the right conditions, these "Lewis acids" are vigorously reactive toward compounds that have electrons to give. In these reactions they form new compounds in which the formerly electron-deficient atoms obtain an octet.

$$BF_3 + :N(CH_3)_3 \longrightarrow F_3\overset{-}{B}-\overset{+}{N}(CH_3)_3$$

There are also many stable compounds in which at least one atom has <u>more</u> than an octet of electrons. The best Lewis structure for the sulfate ion ($SO_4^{=}$) is

$$O=S(=O)(-O^-)_2$$

The phosphate linkage in DNA is written

$$-O-P(=O)(-O^-)-O-$$

In each of these "hypervalent" structures the central atom has more than eight electrons. Hypervalence occurs in the elements of periods 3 to 6; note that these elements have atoms with low-lying d-orbitals that can accomodate additional electrons.

## Exercises

Using the method outlined above, derive the structures for the following compounds. The unbonded skeletons are provided.

18. Propyne ($C_3H_4$)

**Rule 3**

_____ H __________ =

_____ C __________ = _____

_____

**Rule 4 (first trial)**

H

H C C C H

H

**Rule 5 (first trial)**

No. of electrons in structure _____

No. of valence electrons _____

Structure is _____(correct, incorrect)

**(second trial)**

__________ _____ _____

Structure is __________

**(third trial)**

__________ _____ _____

Structure is __________

19. Acetone ($C_3H_6O$)

**Rule 3**

**Rule 4**

H O H

H C C C H

H H

Carry on.

20. Formamide ($CH_3NO$)

**Rule 3**

**Rule 4**

```
    O

H  C  N  H

      H
```

Carry on.

**Rule 3**

**Rule 4**

```
       O
H  N   C   N  H
   H       H
```

21. Urea ($CH_4N_2O$)

22. By now you have discovered that slavishly following rules 1–6 is tedious. As you see and use more and more Lewis structures, frequently you will be able to take shortcuts until writing

```
         H
H        C        H
     C        C
     C        C
H        C        H
         H
```

## *Larger Molecules*

So far, we have written the structures of small molecules. If the procedure outlined were applied scrupulously to writing the Lewis structures of larger molecules, the process would become tiresome. However, organic chemistry is largely the chemistry of functional groups. Often in writing mechanisms or other electron-pushing operations it is only necessary to write the Lewis structure of the functional group. The rest of the molecule can be stipulated in one of several shorthand notations that organic chemists are fond of using. Thus,

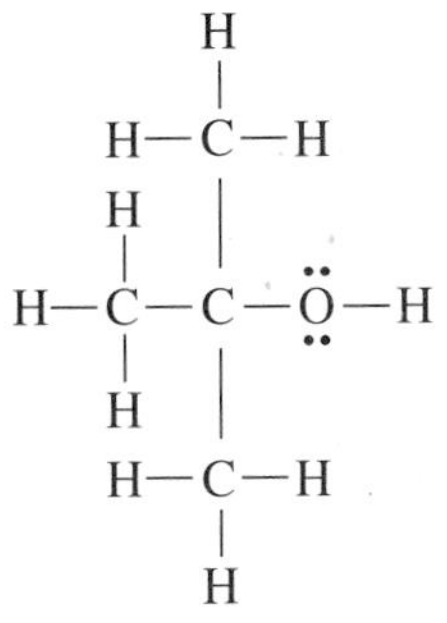

$$H-C(H_2)-C(H)(H)-C(CH_3)-O-H \quad \text{becomes} \quad CH_3-C(CH_3)(CH_3)-O-H$$

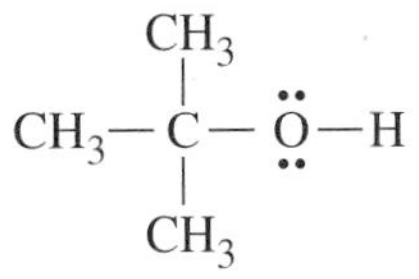

and

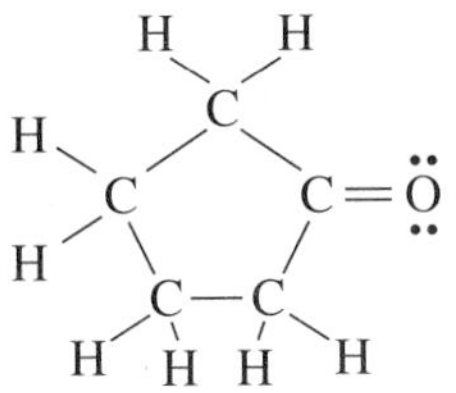

becomes

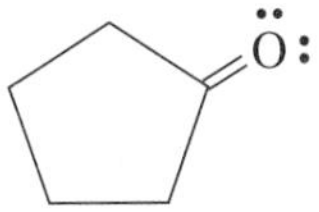

and

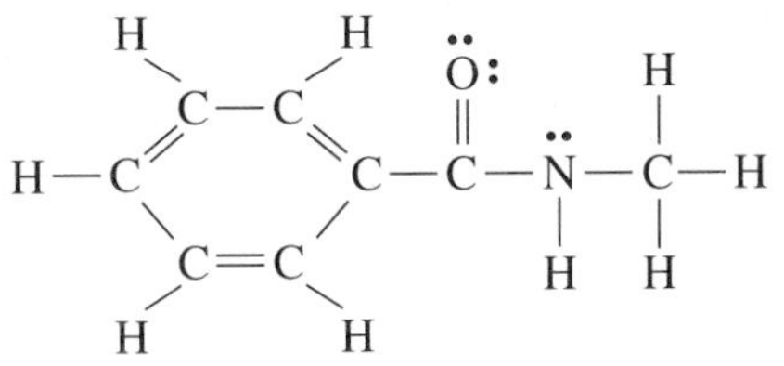

$$H-C-C(=O)-N(H)-C(H)(H)-H \quad \text{becomes} \quad C(=O)-NH-CH_3$$

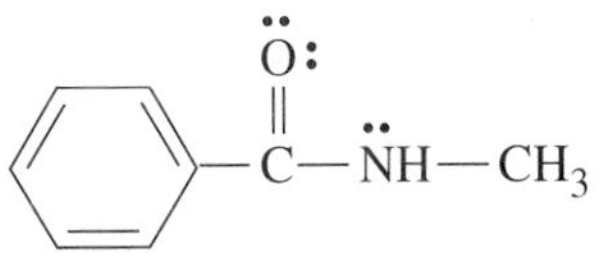

Most organic textbooks do not include the unshared pairs in functional groups. The structures above then become

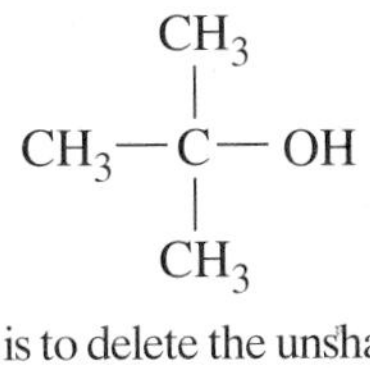

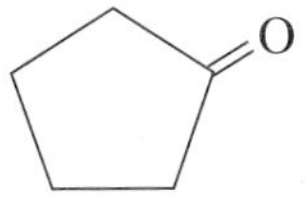

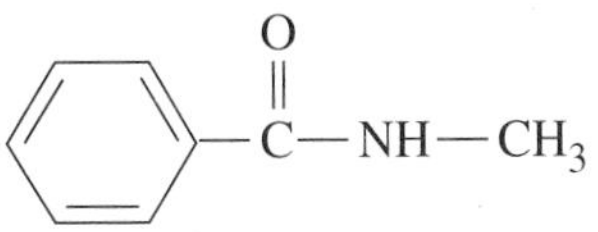

$$CH_3-C(CH_3)(CH_3)-OH \qquad O \qquad C(=O)-NH-CH_3$$

The idea is to delete the unshared pairs and assume that even the beginning student will understand that they must be there because the octet rule applies. This is really a most unfortunate practice because so much of the chemistry of a functional group is determined by the presence (or absence) of unshared pairs.

In order to write Lewis structures of functional groups attached to any alkyl (R) or aryl (Ar) group, one writes down the skeleton (using some shorthand notation) and then counts the available valence electrons allowing each R or Ar attached directly to the functional group to bring one electron into the union.

The use of R or Ar in organic structures seems to confuse some. The letter R is used to denote *any* alkyl (**aliphatic**) group, and Ar is used to denote *any* aryl (**aromatic**) group. This shorthand device allows a chemist to avoid writing a complicated organic structure when all that needs to be shown is the transformation of a functional group. For example,

$$R{-}CH_2{-}\ddot{\underset{\cdot\cdot}{O}}{-}H \xrightarrow{PBr_3} R{-}CH_2{-}\ddot{\underset{\cdot\cdot}{Br}}\!:$$

is much easier than

$$CH_3{-}CH(CH_3){-}CH_2{-}CH(CH_3){-}CH_2{-}\ddot{\underset{\cdot\cdot}{O}}{-}H \xrightarrow{PBr_3} CH_3{-}CH(CH_3){-}CH_2{-}CH(CH_3){-}CH_2{-}\ddot{\underset{\cdot\cdot}{Br}}$$

In this case R =

$$CH_3{-}CH(CH_3){-}CH_2{-}CH(CH_3){-}$$

Likewise

$$Ar{-}C(=\ddot{O}:){-}\ddot{\underset{\cdot\cdot}{O}}{-}H \xrightarrow{SOCl_2} Ar{-}C(=\ddot{O}:){-}Cl$$

is easier than

$$3\text{-}CH_3\text{-}5\text{-}(CH_3{-}\ddot{\underset{\cdot\cdot}{O}}){-}C_6H_3{-}C(=\ddot{O}:){-}\ddot{\underset{\cdot\cdot}{O}}{-}H \xrightarrow{SOCl_2} 3\text{-}CH_3\text{-}5\text{-}(CH_3{-}\ddot{\underset{\cdot\cdot}{O}}){-}C_6H_3{-}C(=\ddot{O}:){-}Cl$$

In this case Ar =

$$3\text{-}CH_3\text{-}5\text{-}(CH_3{-}\ddot{\underset{\cdot\cdot}{O}}){-}C_6H_3{-}$$

23. Phenyl methyl ketone (acetophenone) has the skeleton

$$C_6H_5 \quad \overset{O}{C} \quad CH_3$$

The number of available valence electrons is: from the phenyl group, ___1___; from the methyl group, ___1___ ; from the carbon atom, ___4___; and from the oxygen atom ___6___. The total number of valence electrons is _____. Filling in the skeleton with only single bonds gives

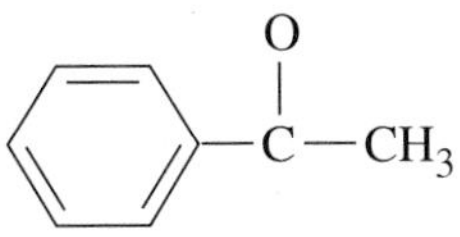

O
|
$C_6H_5$—C—$CH_3$

and adding the appropriate unshared pairs gives

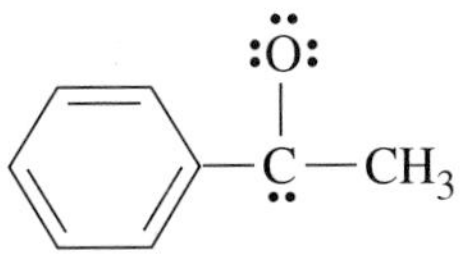

:Ö:
|
$C_6H_5$—C̤—$CH_3$

The number of valence electrons in this structure is _____ which is _____ too many. Removing an unshared pair from carbon and another from oxygen and placing a second bond between carbon and oxygen gives the correct structure:

_______________

24. The skeleton of benzyldimethylamine is

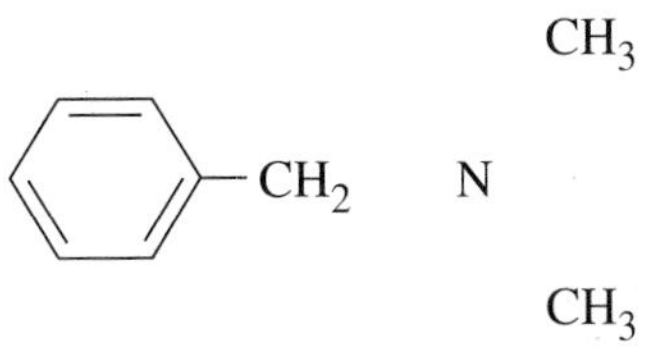

$CH_3$

$C_6H_5$—$CH_2$ N

$CH_3$

The number of available valence electrons is: from the benzyl group, _____; from each of two methyl groups, _____; and, from the nitrogen atom, _____, for a total of _____. Filling in the skeleton with single bonds gives

_______________

and adding the appropriate unshared pair gives

____________________

The number of electrons in the functional group of this structure is _____, and the structure is _____ (correct, incorrect).

25. The skeleton is benzaldoxime is

H
C N
O H

The number of valence electrons is: from the phenyl group, _____, from each of two hydrogens, _____; from the carbon atom, _____; from the nitrogen atom, _____; and from the oxygen atom, _____, for a total of _____. Filling in the skeleton with single bonds and adding the appropriate unshared pairs gives

____________________

The number of electrons in the functional group of this trial structure is _____ which is _____ too many. Removing an unshared pair from carbon and nitrogen and adding a second bond between them gives

____________________

in which carbon, nitrogen, and oxygen have their customary valences of _____, _____, and _____ respectively. The alternative structure with a double bond between nitrogen and oxygen is

____________________

This structure is not acceptable because it requires carbon and oxygen to exhibit the unfamiliar valences of ______ and ______.

## Exercises

Derive Lewis structures for the compounds below.

26. 2-phenyl-2-hexanol

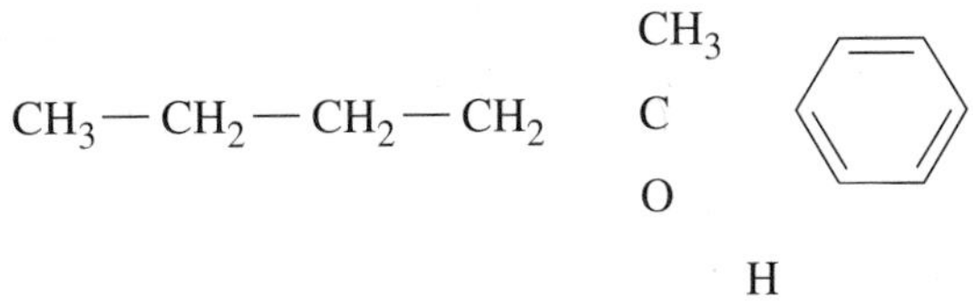

$CH_3$

$CH_3 — CH_2 — CH_2 — CH_2$ C

O

H

27. Furan

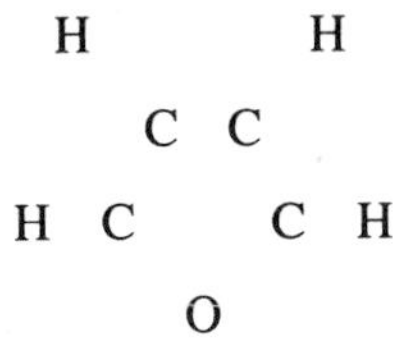

H H

C C

H C C H

O

28. Benzophenone phenylhydrazone

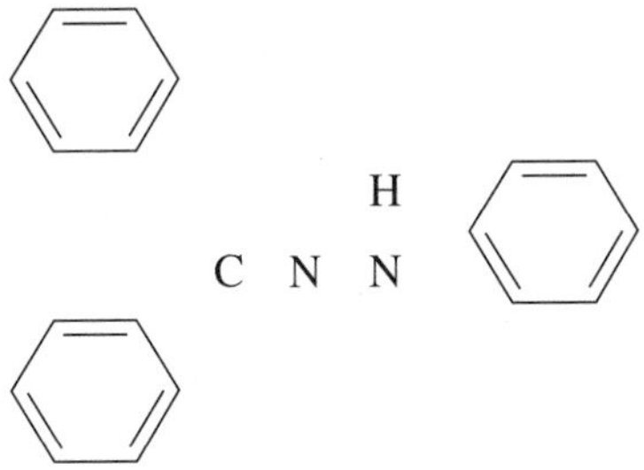

H

C N N

29. Azobenzene

N N

30. Methyl benzimidate

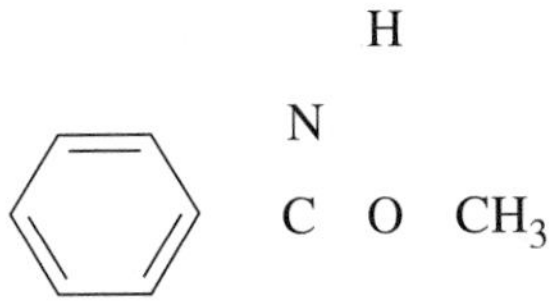

H

N

C O $CH_3$

31. Ethyl crotonate (For the purpose of drawing Lewis structures of compounds with more than one functional group, each group can be treated independently.)

```
        H  H     O
CH3   C  C  C

                 O  CH2—CH3
```

## Formal Charge

Some organic functional groups, although neutral overall, have formal charges on individual atoms. In addition, many of the intermediates that appear in organic reaction mechanisms are charged.

To calculate the **formal charge** on an atom in a particular structure it is necessary to distinguish between the electrons that make up an atom's octet and the electrons that formally "belong" to an atom. The distinction is an arbitrary one, but it is helpful in calculating formal charge. In any Lewis structure *all* electrons associated with an atom either as an unshared pair or in bonding to another atom (shared pair) are part of that atom's octet.

Chloromethane has the Lewis structure

```
     H
     |     ..
H—C—Cl:
     |     ..
     H
```

Circling the carbon atom and its octet of electrons gives

```
     H
   / | \   ..
H(—C—)Cl:
   \ | /   ..
     H
```

Circling the chlorine atom and its octet gives

```
     H
     | / - - \
H—C(— Cl: )
     | \ - - /
     H
```

32. The Lewis structure of acetone is

```
         ..
         O:
         ||
CH3—C—CH3
```

Circling the carbonyl carbon, i.e., the carbon atom attached to oxygen, and its octet gives

```
         ..
         O:
         ||
CH3—C—CH3
```

Circling the oxygen atom and its octet gives

$$\overset{\ddot{O}:}{\overset{\|}{CH_3-C-CH_3}}$$

Thus, atoms share electrons in making bonds, and a pair of electrons may be included in the octet of two different atoms.

When computing the formal charge on an atom, the number of electrons that "belong" to that atom is compared with the number of electrons the atom would have in the unbonded and neutral state. If the two numbers are the same, the formal charge on the atom is zero. In a Lewis structure *both* electrons in an unshared pair belong to the atom, and *one* of every pair of shared (bonding) electrons belongs to the atom.

33. Chloromethane has the Lewis structure

______________

The carbon atom is sharing __4__ electron pairs. In each shared pair the carbon atom "owns" __1__ electron. The number of electrons that "belong" to carbon is ______. Carbon, being a Group ______ element would have __4__ outer shell electrons in the unbonded, neutral state. Therefore, the carbon atom in chloromethane has a formal charge of zero.

34. In the Lewis structure for chloromethane, the chlorine atom is sharing ______electron pair and "owns" ______ of those electrons. Also, the chlorine atom possesses two electrons from each of ______ unshared pairs. The total number of electrons that belong to chlorine is __7__. Chlorine is a Group ______ element. The formal charge on chlorine in chloromethane is ______.

35. The Lewis structure for acetone is

$$\overset{\ddot{O}:}{\overset{\|}{CH_3-C-CH_3}}$$

The carbonyl carbon is sharing ______ pairs of electrons (two carbon–carbon bonds and one carbon–oxygen double bond). From each of those shared pairs the carbonyl carbon "owns" ______ electron. The total number of electrons belonging to the carbonyl carbon is ______. Carbon is a Group ______ element. In acetone the formal charge on the carbonyl carbon is ______.

36. The oxygen atom in acetone possesses ______ unshared pairs and ______ shared pairs of electrons. The number of electrons that belong to oxygen is ______. Oxygen is a Group ______ element. The formal charge on oxygen in acetone is ______.

* * * * *

All of the structures introduced in this chapter so far have only atoms with formal charge equal to zero. We will now see molecules containing atoms with formal charge other than zero.

37. Nitrobenzene has the skeleton

O
N
O

The number of available valence electrons is: from the phenyl group, ______ ; from the nitrogen atom, ______; and from each of two oxygen atoms, ______, for a total of ______. Filling in all single bonds and adding the appropriate unshared pairs gives

________________________________

The **functional group** of this structure contains ______ electrons. Therefore, unshared pairs are removed from the nitrogen atom and one of the oxygen atoms; a double bond is added giving

________________________________

which has the correct number of electrons. In this structure the nitrogen atom is sharing ______ pairs of electrons. From each shared pair the nitrogen owns ______ electron for a total of ______. Nitrogen is a Group ______ element and would have ______ outer shell electrons in the unbonded, neutral state. Since the nitrogen atom in nitrobenzene has one fewer electron than it would in the neutral state, it has a formal charge of ___+1___. This is added to the Lewis structure as + giving

+ N O O

Since an electron is negatively charged, a shortage of one electron results in a single positive charge (+) on an atom. Conversely, an excess of one electron results in a single negative charge (-) on an atom. When an atom in a Lewis structure "owns" two less electrons than it would have in the neutral, unbonded state, it is denoted by ++ or +2 and, conversely, = or -2.

38. Nitrobenzene, a neutral molecule, is not an ion. There must be a formal negative charge somewhere in the molecule to balance the positive charge on the nitrogen. The oxygen atom that is bonded to the nitrogen by a double bond

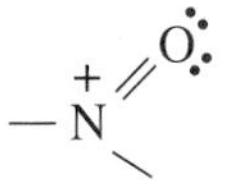

"owns" _____ unshared pairs of electrons. It is sharing _____ pairs of electrons. The number of electrons owned by the oxygen is _____. Since oxygen is a Group _____ element, the formal charge on this oxygen atom is _____. The oxygen atom that is bonded to the nitrogen by a single bond

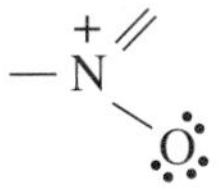

owns _____ unshared pairs of electrons and is sharing _____ pair. The number of electrons owned by this oxygen is _____. This oxygen atom has one more electron than it would have in the neutral state and, thus, has a formal charge of _____. This is added to the Lewis structure as – giving

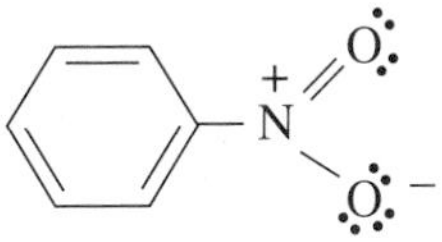

which is a complete, correct Lewis structure.

The cognoscenti in the crowd will recognize that this structure is one of four equivalent resonance structures for nitrobenzene. This will come up later in the text.

## Exercises

Compute and add on the formal charges in these Lewis structures.

39. Pyridine N-oxide

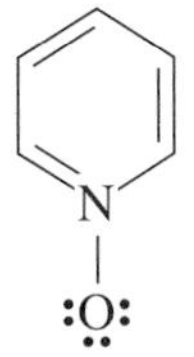

40. Benzenesulfonic acid

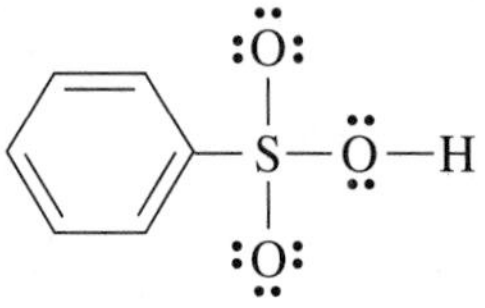

41. N-methyl benzenesulfonamide

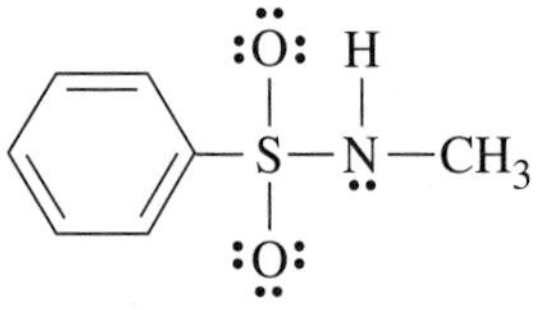

42. Ethylidenetriphenylphosphorane (an ylide)

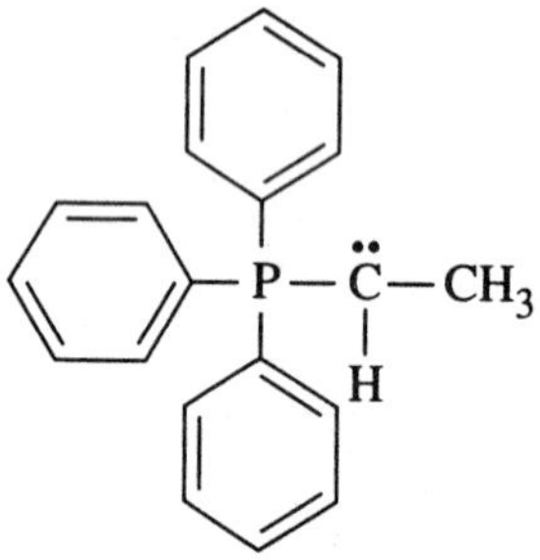

43. Methylazide

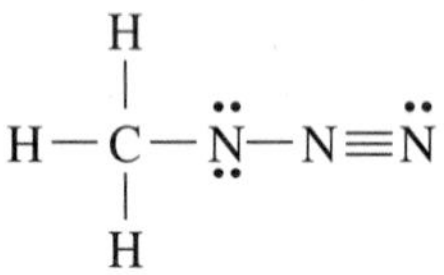

44. Diazomethane

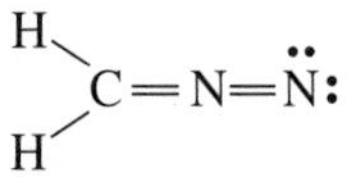

45. Phenylisocyanide

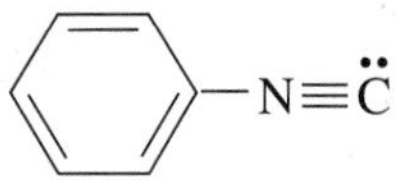

46. Phenylcyanide (benzonitrile)

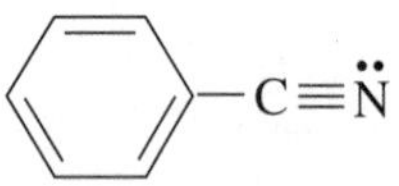

47. Trimethylamine oxide

$$\mathrm{CH_3-\overset{\displaystyle CH_3}{\underset{\displaystyle CH_3}{N}}-\ddot{\underset{\cdot\cdot}{O}}:}$$

48. Dimethylsulfoxide

$$\mathrm{CH_3-\overset{\displaystyle :\ddot{O}:}{\underset{\cdot\cdot}{S}}-CH_3}$$

## Ions

The structures encountered so far have been those of molecules, i.e., neutral species. In many cases there was no formal charge anywhere in the structure. In some cases formal charges appeared in the structure, but their algebraic sum was zero.

Although the starting materials and products of most organic reactions are molecules, a very large number of organic reactions involve ions as intermediates. Ions are charged species. The algebraic sum of the formal charges in an ion is not zero. With very few exceptions the ions encountered in organic chemistry will have a total charge of +1 or -1. Ions with a total charge of greater than one, which are quite common in inorganic chemistry, are rarely found in organic chemistry.

The ionic intermediates of organic chemistry are usually short-lived species because they are unstable relative to neutral molecules. Nevertheless, the proposed existence of ionic intermediates has contributed enormously to understanding organic reactions.

The most direct way to comprehend the structures of ions is to see how they arise in a reaction. This involves some sort of bond-breaking or bond-making process. Bond breaking and bond making are covered extensively in Chapter 3. They will be introduced briefly here.

### Cations

Consider a molecule consisting of a methyl group attached to chlorine (chloromethane)

$$\mathrm{H-\overset{\displaystyle H}{\underset{\displaystyle H}{C}}-\ddot{\underset{\cdot\cdot}{Cl}}:}$$

The formal charges on all the atoms in this molecule are zero. Now consider what would result if this molecule were broken at the C–Cl bond so that the two electrons that constitute the C–Cl bond both depart with the chlorine.

$$\mathrm{H-\overset{\displaystyle H}{\underset{\displaystyle H}{C}}-\ddot{\underset{\cdot\cdot}{Cl}}:} \longrightarrow \mathrm{H-\overset{\displaystyle H}{\underset{\displaystyle H}{C}}\qquad :\ddot{\underset{\cdot\cdot}{Cl}}:}$$

### A Digression

We have a useful notation for the process. It uses curved arrows to indicate the movement of electrons. When organic chemists use the method, they are accused of "pushing electrons." The notation

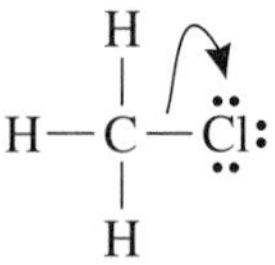

$$H_3C{-}\ddot{\underset{\cdot\cdot}{Cl}}{:}$$

exactly illustrates chloromethane's being "broken at the C–Cl bond so that the two electrons that constitute the C–Cl bond both depart with the chlorine."

Carefully note these three points:

1. The base of the arrow begins *at the original location of the pair of electrons* (in this case, the C–Cl bond).
2. The head of the arrow points *to the destination* of the electrons (the chlorine atom).
3. When the head of the arrow has *two* barbs, it denotes the movement of a *pair* of electrons.

49. The usual method is used to calculate the formal charge on the resulting fragments. The carbon atom of the methyl fragment is sharing ____ electron pairs. In each shared pair the carbon "owns" _____ electron. The number of electrons which "belong" to carbon is _____. Carbon, being a Group _____ element would have _____ outer shell electrons in the neutral, unbonded state. Since the carbon atom has one fewer electron than it would in the neutral state it has a formal charge of _____. Therefore, the correct Lewis structure for the methyl cation is

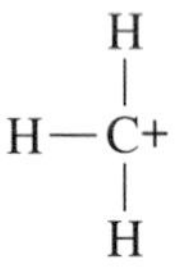

$$H_3C^{+}$$

The chlorine fragment now possesses _____ unshared electron pairs. The number of electrons which belong to chlorine is _____. This Group VII element has a formal charge of _____ and has become the familiar chloride ion. Note the conservation of charge. In a chemistry equation the algebraic sum of the charges on one side *must* equal the sum on the other side. Thus, in this example, a neutral molecule (charge = 0) yields a cation plus an anion [charge =(+1) + (-1) = 0].

The arrow provides a strong indication of what formal charges will result. If a neutral molecule is cleaved, an excess of electrons (negative charge) will result at the head of the arrow and a deficiency of electrons (positive charge) will result at the tail of the arrow.

The reaction,

$$CH_3{-}\ddot{\underset{\cdot\cdot}{Cl}}{:} \longrightarrow CH_3^{+} + {:}\ddot{\underset{\cdot\cdot}{Cl}}{:}^{-}$$

provides an example where the two electrons of the C–Cl bond both went with the chlorine. Two ions, one electron-poor and the other electron-rich, result from this **heterolytic** (unsymmetrical) bond **cleavage**.

50. The *n*-propyl cation can be formed from a molecule such as

$$CH_3-CH_2-CH_2-\ddot{\underset{\cdot\cdot}{Cl}}:$$

When the C–Cl bond is broken so that both electrons leave with Cl, the fragments formed are

$$\_\_\_\_\_\_\_\_\_\_ \quad + \quad :\ddot{\underset{\cdot\cdot}{Cl}}:$$

The carbon atom that had been attached to Cl is now sharing ______ electron pairs. In each shared pair the carbon atom owns ______ electron. The number of electrons which belong to carbon is ______. The formal charge on the carbon atom is ______. The correct Lewis structure for the *n*-propyl cation is

$$\_\_\_\_\_\_\_\_\_\_ \quad + \quad :\ddot{\underset{\cdot\cdot}{Cl}}:^{-}$$

51. The isopropyl cation can be formed from

$$CH_3-\overset{\displaystyle H}{\underset{\displaystyle CH_3}{\overset{|}{\underset{|}{C}}}}-\ddot{\underset{\cdot\cdot}{Br}}:$$

When the C–Br bond is broken so that both electrons go with Br (bromine departs as bromide ion), the organic fragment formed is

$$\_\_\_\_\_\_\_\_\_\_$$

The carbon atom that had been attached to bromine is now sharing ______ electron pairs. The total number of electrons that belong to carbon is ______. The formal charge on the carbon atom is ______. The correct Lewis structure for the isopropyl cation is

______________ + $:\ddot{\underset{..}{Br}}:^-$

52. The cyclopentyl cation can be formed from

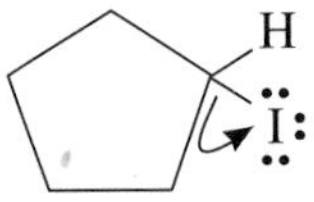

(cyclopentyl iodide structure: cyclopentane ring bearing H and $:\ddot{\underset{..}{I}}:$, with a curved arrow from the C–I bond toward iodine)

When the iodine departs as iodide ion the organic fragment formed is

______________

The carbon atom that had been attached to iodine is now sharing ______ electron pairs. The total number of electrons belonging to the carbon atom is ______. The formal charge on the carbon atom is ______. The correct Lewis structure for the cyclopentyl cation is

______________ + $:\ddot{\underset{..}{I}}:^-$

53. Methanol, $CH_3$—O—H, is a compound in which the formal charge on all the atoms is zero. Consider what results when a proton, $H^+$, becomes bonded to methanol by way of one of the unshared electron pairs on the oxygen atom, i.e.,

$$CH_3-\ddot{O}:(-H) + H^+ \longrightarrow CH_3-\ddot{O}(-H)(-H)$$

This time we use the curved arrow to signify bond making. Now a pair of unshared electrons on oxygen is pushed toward the region between the oxygen atom and the hydrogen ion. It becomes an O–H covalent bond.

In the resulting structure the oxygen atom owns one electron from each of ______ shared pairs and two electrons from ______ unshared pair. The total number of electrons which belong to the oxygen atom is ____. Oxygen is a Group ____ element. Since the number of electrons which the oxygen atom owns in this structure is one less than it would have in the neutral, unbonded state, the charge on oxygen is ____. The correct Lewis structure for the conjugate acid of methanol is

______________________

Once again charge is conserved. Joining a neutral molecule and a cation *must* yield a cation. Again the formal charge distribution on the resulting ion is predictable from the arrow. Electrons are pushed *away* from the oxygen atom, leaving it with a positive charge. Electrons are pushed *toward* the hydrogen ion, neutralizing its erstwhile positive charge.

54. When a proton becomes bonded to diethyl ether, $CH_3-CH_2-\ddot{\underset{\cdot\cdot}{O}}-CH_2-CH_3$, by way of one of the unshared electron pairs on the oxygen atom, the result is

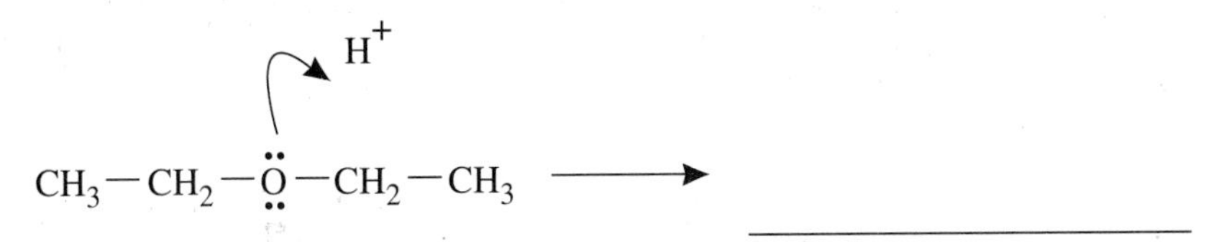


In this structure the oxygen atom owns one electron from each of ____ shared pairs and two electrons from ____ unshared pair. The total number of electrons that belong to oxygen is ____. The formal charge on oxygen is ____. The correct Lewis structure for the conjugate acid of diethyl ether is

______________________

55. Tetrahydrofuran has the structure,

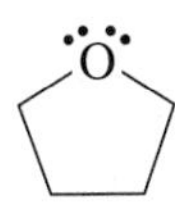


When a proton becomes bonded to tetrahydrofuran the result is

______________________

In this structure the oxygen atom owns one electron from each of _____ shared pairs and two electrons from _____ unshared pair. The formal charge on the oxygen atom is _____. The Lewis structure for the conjugate acid of tetrahydrofuran is

_____________________

56. Methylamine has an unshared electron pair on the nitrogen atom. When a proton becomes bonded to methylamine, the result is

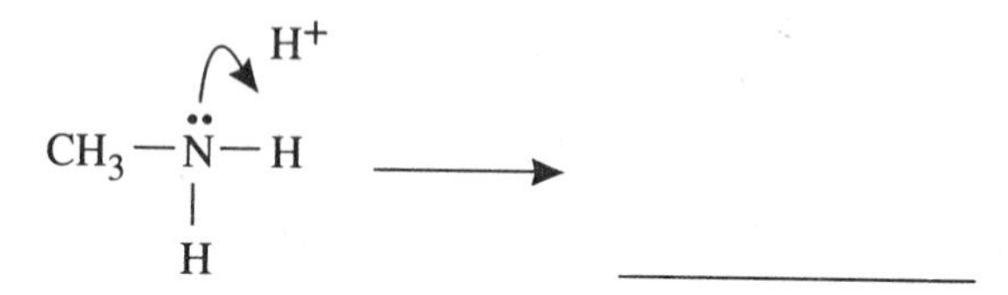


In this structure the nitrogen atom is sharing _____ electron pairs. From each shared pair the nitrogen owns _____ electron for a total of _____ electrons. Nitrogen is a Group _____ element. The formal charge on the nitrogen atom is _____. The Lewis structure for the methylammonium ion is

_____________________

57. When a proton becomes bonded to diethylamine the result is

$CH_3—CH_2$ $H^+$
N
$CH_3—CH_2$ H
→ _____________________

In this structure the nitrogen owns one electron from each of _____ shared pairs. The total number of electrons which belong to nitrogen is _____. The formal charge on the nitrogen is _____. The Lewis structure for the diethylammonium ion is

____________________

58. The structure of pyridine is

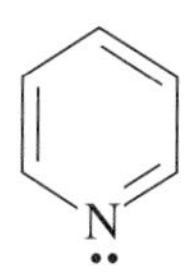

When a proton becomes bonded to the nitrogen atom by way of its unshared electron pair, the result is

____________________

In this structure the nitrogen atom is sharing _____ electron pairs. The formal charge on the nitrogen atom is _____. The Lewis structure for the pyridinium ion is

____________________

## *Anions*

In methyl lithium the carbon atom is bound to a metal atom. The formal charges on all the atoms in the molecule are zero. Consider the result if the C–Li bond were broken so that the two electrons that constitute the bond both *remain* with the carbon.

```
     H                      H
     |↶                     |
H — C — Li   ———►    H — C:          +   Li
     |                      |
     H                      H
```

59. The carbon atom owns one electron from each of _____ shared pairs and two electrons from

_____ unshared pair. The number of electrons that belong to carbon is _______. Carbon is a Group _____ element. Since the carbon atom has one more electron than it would in the neutral, unbonded state, it has a formal charge if -1. The Lewis structure for the methyl anion is

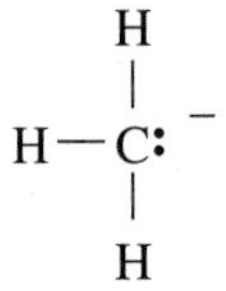

$$H-\overset{\overset{\displaystyle H}{|}}{\underset{\underset{\displaystyle H}{|}}{C}}{:}^{-}$$

The lithium fragment must have a formal charge of _____.

60. The *n*-butyl anion can be formed from

$$CH_3-CH_2-CH_2-\overset{\overset{\displaystyle H}{|}}{\underset{\underset{\displaystyle H}{|}}{C}}-Li$$

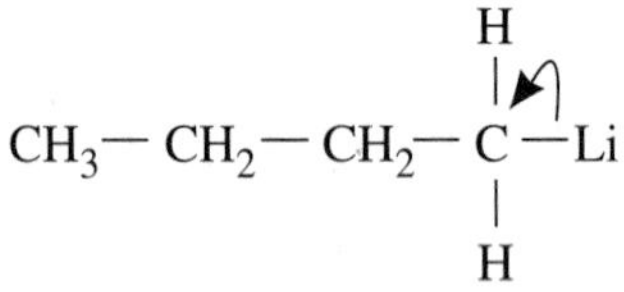

When the C–Li bond is broken so that both electrons remain with the carbon atom, the organic fragment formed is

_______________

The carbon atom that had been attached to Li now owns one electron from each of _____ shared pairs and two electrons from _____ unshared pair. The number of electrons that belong to carbon is _____. The formal charge on the carbon atom is _____. The Lewis structure for the *n*-butyl anion is

+ $Li^+$

_______________

61. The isobutyl anion can be formed from

$$CH_3-\underset{\underset{\displaystyle CH_3}{|}}{CH}-\overset{\overset{\displaystyle H}{|}}{\underset{\underset{\displaystyle H}{|}}{C}}-Na$$

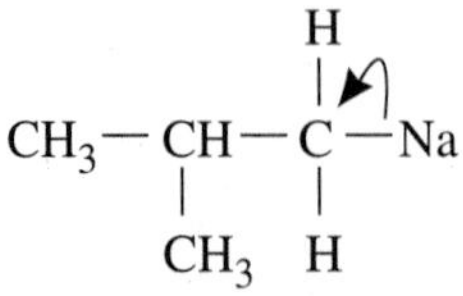

When the C–Na bond is broken so that both electrons remain with the carbon atom, the organic fragment formed is

The carbon atom that had been attached to Na now owns one electron from each of _____ shared pairs and two electrons from _____ unshared pair. The number of electrons that belong to carbon is _____. The Lewis structure for the isobutyl anion is

$+ \quad Na^+$

---

62. When the C–Mg bond in

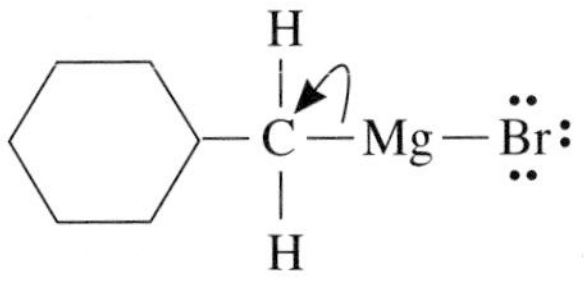

$$\text{C}_6\text{H}_{11}-\underset{\text{H}}{\overset{\text{H}}{\text{C}}}-\text{Mg}-\ddot{\underset{\cdot\cdot}{\text{Br}}}\text{:}$$

is broken so that both electrons remain with the carbon atom, the fragment formed is

The carbon atom that had been attached to Mg now owns one electron from each of _____ shared pairs and two electrons from _____ unshared pair. The formal charge on the carbon atom is _____. The Lewis structure for the cyclohexylmethyl anion is

$+ \quad {}^+MgBr$

---

The main player in organic chemistry is carbon. It follows that organic chemists concentrate on the fate of carbon in these reactions. Compare the cleavage reactions (59–62) with (49–52). All are heterolytic cleavages. In 59–62 the electrons move toward the carbon, generating a **carbanion.** Conversely, in 49–52 the electrons move away from carbon and produce a **carbocation**. The original polarity of the bond influences the direction of heterolysis; the more electronegative atom takes the electrons. Thus, the electrons go with the carbon when a carbon–metal bond breaks, but go with the halogen when a carbon–halogen bond breaks.

63. Ethanol, $CH_3-CH_2-\ddot{\underset{\cdot\cdot}{O}}-H$, is a compound in which the formal charge on all the atoms is zero. Under certain conditions the O–H bond can be broken so that both electrons remain with the oxygen atom. The resulting fragments are

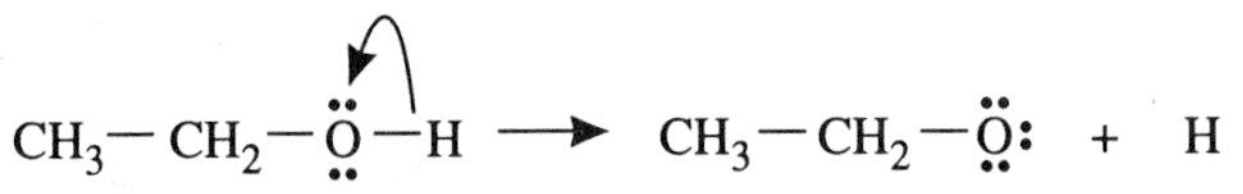

$$CH_3-CH_2-\ddot{\underset{\cdot\cdot}{O}}-H \longrightarrow CH_3-CH_2-\ddot{\underset{\cdot\cdot}{O}}: \; + \; H$$

In this structure the oxygen owns one electron from _____ shared pair and two electrons from each of _____ unshared pairs. The total number of electrons belonging to oxygen is _____. Oxygen is a Group _____ element. The formal charge on the oxygen atom is _____. The correct Lewis structure for the ethoxide ion is

_______________ + $H^+$

Note that the other fragment, the proton, leaves with a formal charge of +1.

64. The loss of a proton attached to the oxygen atom of *t*-butyl alcohol results in

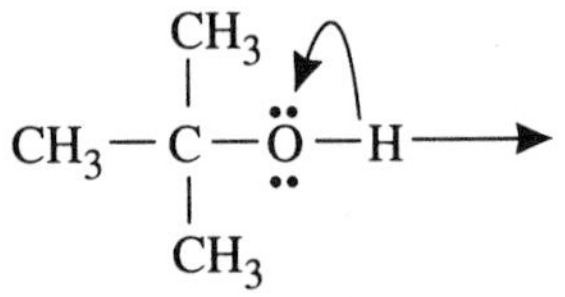

$$CH_3-\underset{CH_3}{\overset{CH_3}{\underset{|}{\overset{|}{C}}}}-\ddot{\underset{\cdot\cdot}{O}}-H \longrightarrow \_\_\_\_\_\_\_\_\_\_\_\_ + \; H$$

In this structure the oxygen atom owns one electron from _____ shared pair and two electrons from each of _____ unshared pairs. The number of electrons belonging to oxygen is _____. The formal charge on the oxygen is _____. The Lewis structure for the *t*-butoxide ion is

_______________ + $H^+$

65. A very strong base can remove a proton from methylamine,

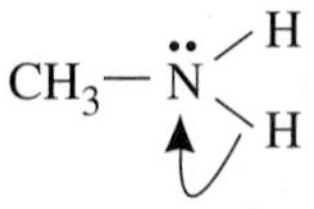

$$CH_3-\ddot{N}\begin{matrix} H \\ H \end{matrix}$$

The proton removed is one of those attached to the nitrogen atom. The resulting fragments are

+ H

______________

The nitrogen atom now owns one electron from each of _____ shared pairs and two electrons from each of _____ unshared pairs. The number of electrons that belong to the nitrogen atom is _____. Nitrogen is a Group _____ element. The formal charge on nitrogen is _____. The Lewis structure for the methylamide ion is

+ $H^+$

______________

66. The loss of a proton from diisopropylamine,

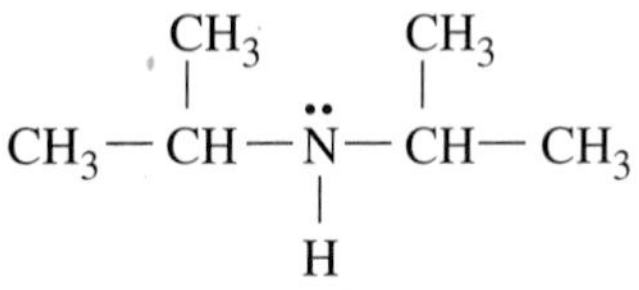

$$CH_3-\underset{\displaystyle CH_3}{CH}-\ddot{N}(H)-\underset{\displaystyle CH_3}{CH}-CH_3$$

results in the fragment

______________

The number of electrons which belong to the nitrogen atom is _____. The formal charge on the nitrogen atom is _____. The correct Lewis structure for the diisopropylamide ion is

+ $H^+$

______________

# Free Radicals

Covalent bonds can suffer *homolytic* (symmetrical) bond breaking. During homolytic cleavage, the electrons of a covalent bond become unpaired and one electron goes with each fragment. Radicals, rather than ions, are formed. Radicals usually have a formal charge of zero.

Consider the homolytic cleavage of a $Cl_2$ molecule.

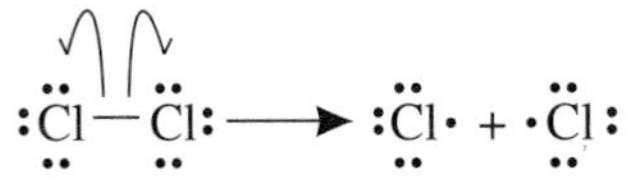

$$:\ddot{\underset{\cdot\cdot}{Cl}}-\ddot{\underset{\cdot\cdot}{Cl}}: \longrightarrow :\ddot{\underset{\cdot\cdot}{Cl}}\cdot + \cdot\ddot{\underset{\cdot\cdot}{Cl}}:$$

The entirely symmetrical process yields two chlorine atoms. Since each chlorine "owns" seven electrons, the formal charge is zero. The presence of an *unpaired* electron identifies each fragment as a "free radical."

Note carefully how the curved arrow notation has changed! The base of the arrow still begins at the original location of the electron and the arrow's head still points to the destination of the electron. However, now the head of the arrow has *only one barb* signifying the movement of a single electron. Two such arrows are required to denote homolytic bond breaking since each electron goes its separate way.

67. When a $Br_2$ molecule undergoes homolytic cleavage, it yields ______ (number) bromine radicals. Thus,

$$:\ddot{\underset{\cdot\cdot}{Br}}-\ddot{\underset{\cdot\cdot}{Br}}: \longrightarrow \underline{\qquad} + \underline{\qquad}$$

Each radical has a formal charge of zero.

Homolytic bond breaking usually requires quite large energies. The homolysis of $Cl_2$ or $Br_2$ is induced only at high temperature or by ultraviolet light. Many free radical reactions are promoted photochemically. Homolytic cleavage reactions important to organic chemistry often occur when the covalent bond in question is weak. Oxygen–oxygen (peroxide) bonds are a good example.

68. When di-*tert*-butyl peroxide breaks homolytically, it forms two identical radicals with the Lewis structure

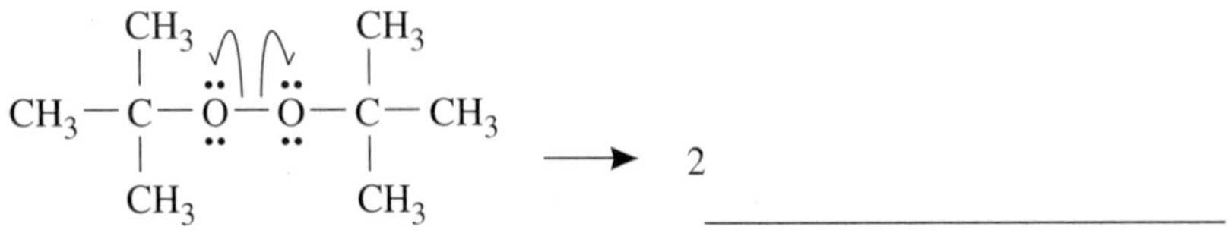

$$(CH_3)_3C-\ddot{\underset{\cdot\cdot}{O}}-\ddot{\underset{\cdot\cdot}{O}}-C(CH_3)_3 \longrightarrow 2\ \underline{\qquad\qquad\qquad}$$

The oxygen atom owns ______ electron from a shared pair, ______ unshared electrons for a total of _____ electrons. Oxygen is a Group VI element, and its formal charge is _____.

69. The homolysis of the O–O bond in diacetyl peroxide gives two acetyl radicals with Lewis structure

$$CH_3-\overset{\overset{\displaystyle :\ddot{O}}{\|}}{C}-\ddot{\underset{\cdot\cdot}{O}}-\ddot{\underset{\cdot\cdot}{O}}-\overset{\overset{\displaystyle \ddot{O}:}{\|}}{C}-CH_3 \longrightarrow 2\ \_\_\_\_\_\_\_\_\_\_$$

The formal charge on this radical is _____.

70. Homolysis of an unsymmetrical bond will result in two different radicals. The homolytic cleavage of the O–O bond in *tert*-butyl hydroperoxide will yield

$$CH_3-\underset{\underset{\displaystyle CH_3}{|}}{\overset{\overset{\displaystyle CH_3}{|}}{C}}-\ddot{\underset{\cdot\cdot}{O}}-\ddot{\underset{\cdot\cdot}{O}}-H \longrightarrow \_\_\_\_\_\_\_\_\_\_ + \cdot\ddot{\underset{\cdot\cdot}{O}}-H$$

71. Homolysis of *m*-chloroperbenzoic acid will give

$$m\text{-}(:\ddot{\underset{\cdot\cdot}{Cl}}:)C_6H_4-\overset{\overset{\displaystyle \ddot{O}:}{\|}}{C}-\ddot{\underset{\cdot\cdot}{O}}-\ddot{\underset{\cdot\cdot}{O}}-H \longrightarrow \_\_\_\_\_\_\_\_\_\_ + \_\_\_\_\_\_\_\_\_\_$$

72. Carbon radicals are usually produced indirectly. An exception is the homolytic cleavage of azobis(isobutyronitrile). In this commercially important reaction, the C–N bonds break homolytically to give dinitrogen plus two identical carbon radicals.

$$CH_3-\underset{\underset{\displaystyle CH_3}{|}}{\overset{\overset{\displaystyle CN}{|}}{C}}-\ddot{N}=\ddot{N}-\underset{\underset{\displaystyle CH_3}{|}}{\overset{\overset{\displaystyle CN}{|}}{C}}-CH_3 \longrightarrow 2\ \_\_\_\_\_\_\_\_\_\_ + \ddot{N}\equiv\ddot{N}$$

Reaction conditions dictate how molecules cleave. Write the products from isopropyl hydroperoxide in each of the reactions described.

73. *Hetero*lytic cleavage of the O–H bond to release a proton.

$(CH_3)_2CH-\ddot{O}-\ddot{O}-H \longrightarrow$ ________________ + $H^+$

74. *Homo*lytic cleavage of the O–O bond.

$(CH_3)_2CH-\ddot{O}-\ddot{O}-H \longrightarrow$ ________________ + ________________

75. *Hetero*lytic cleavage of the C–O bond to yield a carbocation.

$(CH_3)_2CH-\ddot{O}-\ddot{O}-H \longrightarrow$ ________________ + ________________

# Answers

## Chapter 1

1. IV, 4, 6

2. 1, 4, 6, 14

| | | | |
|---|---|---|---|
| 4 | H | $1 \times 4 =$ | 4 |
| 1 | C | $4 \times 1 =$ | 4 |
| 1 | O | $6 \times 1 =$ | 6 |
| | | | 14 |

3.
```
   H
H  C  Cl
   H
```

4.
```
   H                 H
                     |
H  C  O  H  ,  H—C—O—H
                     |
   H                 H
```

```
     H
     |    ••
H—C—O—H
     |    ••
     H
```

5. 14

6. 14, 14, correct

7.

| | | | |
|---|---|---|---|
| 6 | H | $1 \times 6 =$ | 6 |
| 2 | C | $4 \times 2 =$ | 8 |
| 1 | O | $6 \times 1 =$ | 6 |
| | | | 20 |

```
     H          H
     |    ••    |
H—C—O—C—H   , 20, 20, correct
     |    ••    |
     H          H
```

8. 5 H $1 \times 5 = 5$ ,
1 C $4 \times 1 = 4$
1 N $5 \times 1 = 5$
14

H—C(H)(H)—N̈(H)—H , 14, 14, correct

9. 4 H $1 \times 4 = 4$ ,
1 C $4 \times 1 = 4$
1 S $6 \times 1 = 6$
14

H—C(H)(H)—S̈—H , 14, 14, correct

10. 8 H $1 \times 8 = 8$ ,
3 C $4 \times 3 = 12$
2 O $6 \times 2 = 12$
32

H—C(H)(H)—Ö—C(H)(H)—Ö—C(H)(H)—H , 32, 32, correct

11. 1, 2, 4, 12

12. 14

13. 12

14. 2, 1, 4, 6, 12, 2 H $1 \times 2 = 2$ ,
1 C $4 \times 1 = 4$
1 O $6 \times 1 = 6$
12

(H)(H)C—O, (H)(H)C̈—Ö:,

14, incorrect, (H)(H)C=Ö: , 12, correct

15. 3, 1, 2, 4, 5, 16, 3 H $1 \times 3 = 3$ ,
2 C $4 \times 2 = 8$
1 N $5 \times 1 = 5$
16

H—C(H)(H)—C—N,

H—C(H)(H)—C̈—N̈: , 20, incorrect, H—C(H)(H)—C̈=N̈: , 18, incorrect,

H—C(H)(H)—C≡N: , 16, correct,

16. 18, 2 H $1 \times 2 = 2$ , 20
1 C $4 \times 1 = 4$
2 O $6 \times 2 = 12$
18

17.

```
     H      ..
     |    //O:
  H—C—C
     |     \ ..
     H       Cl:
              ..
```

18.

4 H $1 \times 4 =$ 4
3 C $4 \times 3 =$ 12
16

```
     H
     |   ..  ..
  H—C—C—C—H
     |   ..  ..
     H
```

20
16
incorrect

```
     H
     |
  H—C—C=C—H
     |   ..  ..
     H
```

18
16
incorrect

```
     H
     |
  H—C—C≡C—H
     |
     H
```

16
16
correct

19.

6 H $1 \times 6 =$ 6
3 C $4 \times 3 =$ 12
1 O $6 \times 1 =$ 6
24

```
     H  :Ö:  H
     |   |   |
  H—C—C—C—H
     |   ..  |
     H       H
```

26
24
incorrect

```
     H  Ö:  H
     |  ||  |
  H—C—C—C—H
     |       |
     H       H
```

24
24
correct

20.

3 H $1 \times 3 =$ 3
1 C $4 \times 1 =$ 4
1 O $6 \times 1 =$ 6
1 N $5 \times 1 =$ 5
18

```
      :Ö:
       |   ..
  H—C—N—H
      ..  |
          H
```

20
18
incorrect

```
     Ö:
     ||  ..
  H—C—N—H
          |
          H
```

18
18
correct

(alternative structure,

```
      :Ö:
       |
  H—C=N—H
          |
          H
```

not as good because oxygen has a valence of one)

21.

| | | | |
|---|---|---|---|
| 4 | H | $1 \times 4 =$ | 4 |
| 1 | C | $4 \times 1 =$ | 4 |
| 1 | O | $6 \times 1 =$ | 6 |
| 2 | N | $5 \times 2 =$ | 10 |
| | | | 24 |

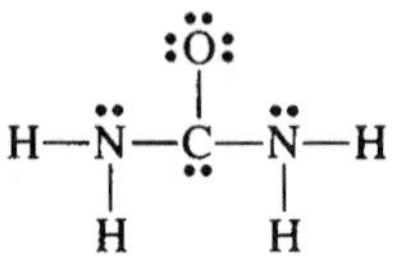

, [structure], , 26 24 incorrect

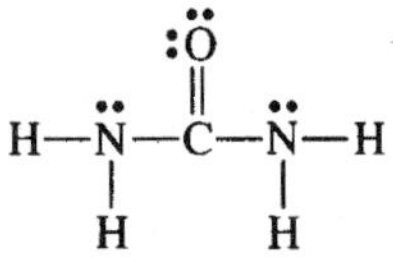

[structure], 24 24 correct , (alternative structure, [structure] is not as good because oxygen has a valence of one)

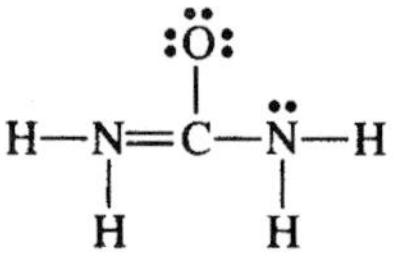

22. 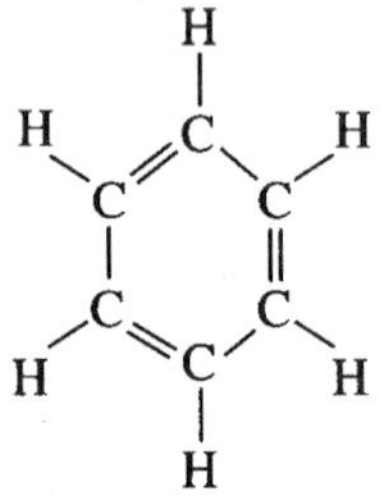 [structure] or [structure]

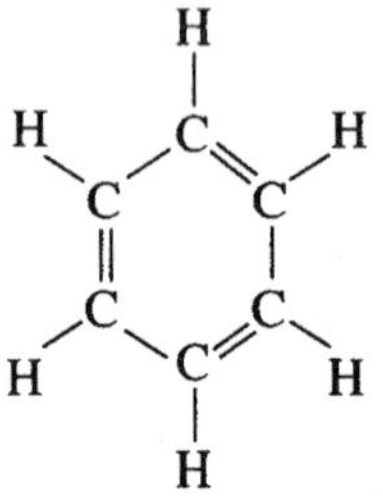

23. 12, 14, 2, [structure]

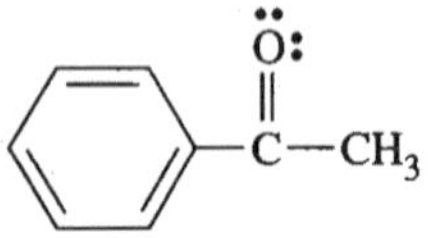

24. 1, 1, 5, 8, [structure] , [structure] , 8, correct

25. 1, 1, 4, 5, 6, 18, [structure] , 20, 2,

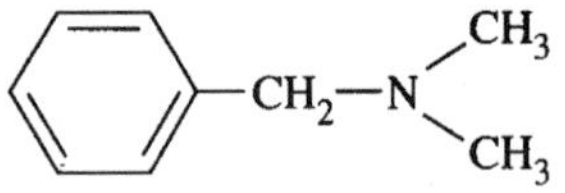

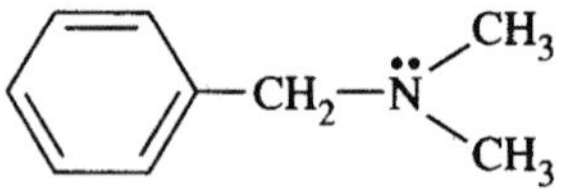

[structure] , 4, 3, 2, [structure] , 3, 3

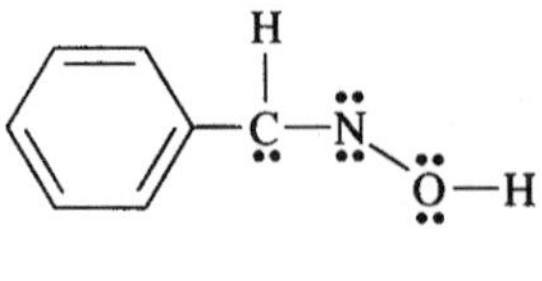

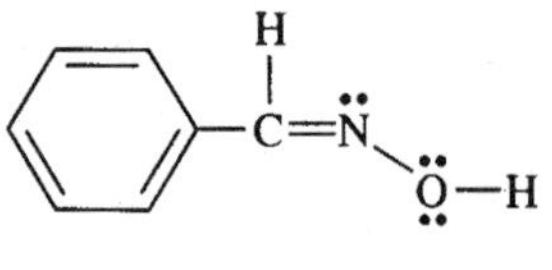

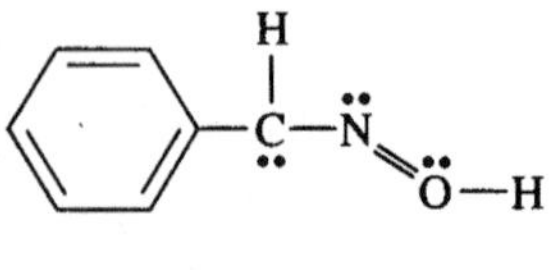

26.

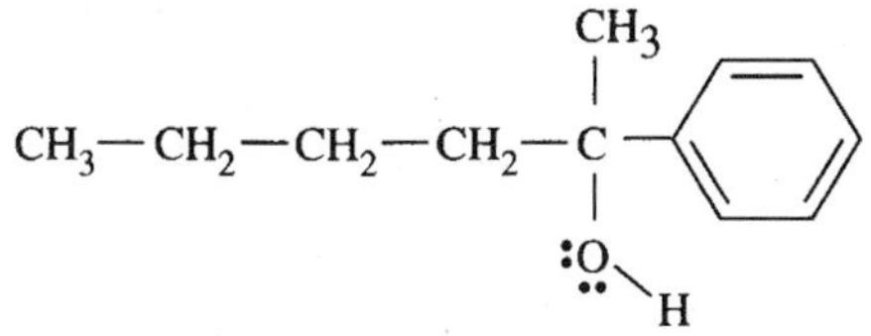

$CH_3$

$CH_3-CH_2-CH_2-CH_2-C-$

$:\ddot{O}-H$

27.

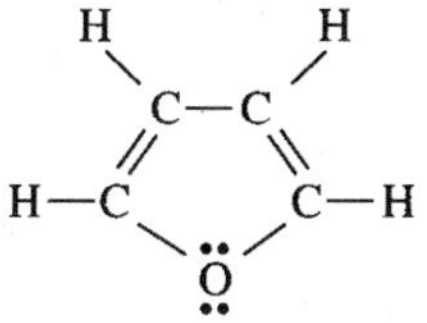

H H

C—C

H—C C—H

$\ddot{\underset{\cdot\cdot}{O}}$

28.

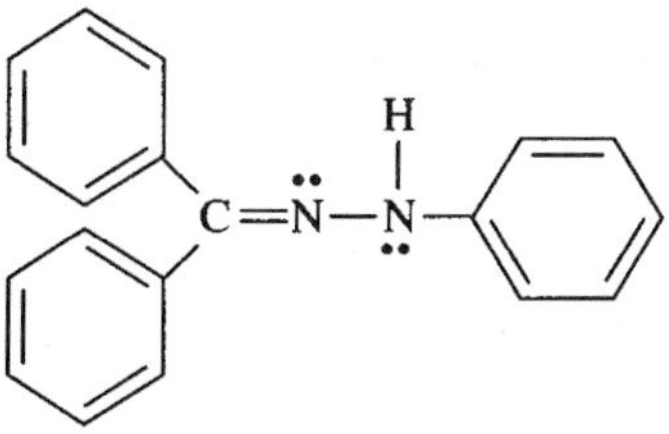

H

$C=\ddot{N}-\underset{\cdot\cdot}{N}-$

29.

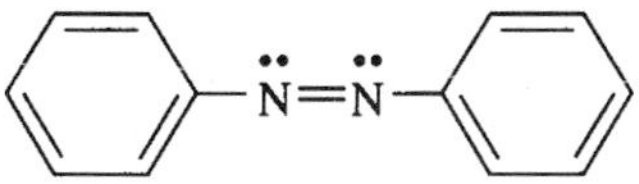

$-\ddot{N}=\ddot{N}-$

30.

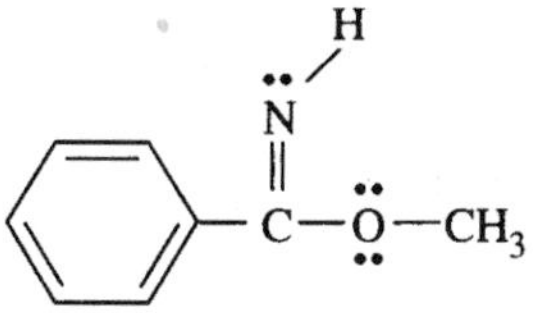

H

$\ddot{N}$

$C-\ddot{\underset{\cdot\cdot}{O}}-CH_3$

31.

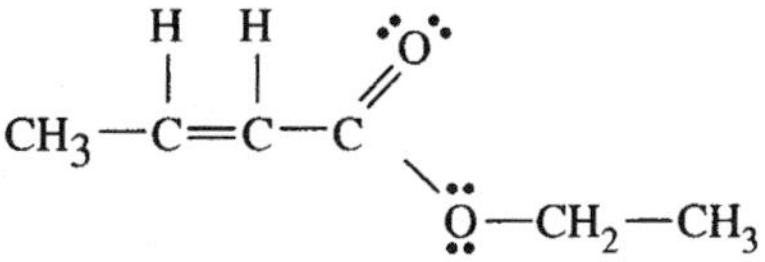

H H $\ddot{O}$

$CH_3-C=C-C$

$\ddot{\underset{\cdot\cdot}{O}}-CH_2-CH_3$

32.

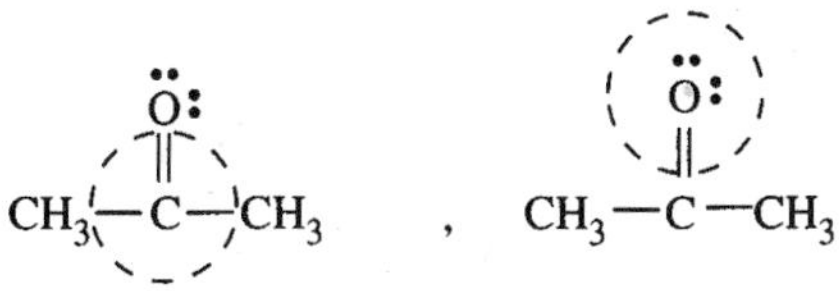

$\ddot{O}:$

$CH_3-C-CH_3$ , $CH_3-C-CH_3$

33.

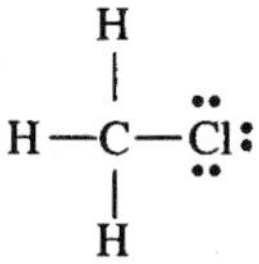

H

$H-C-\ddot{\underset{\cdot\cdot}{Cl}}:$ , 4, IV

H

34. 1, 1, 3, VII, zero

35. 4, 1, 4, IV, zero

36. 2, 2, 6, VI, zero

37. 1, 5, 6, 18, 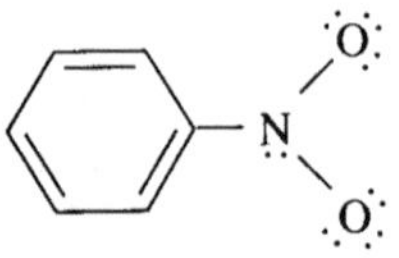 , 20, 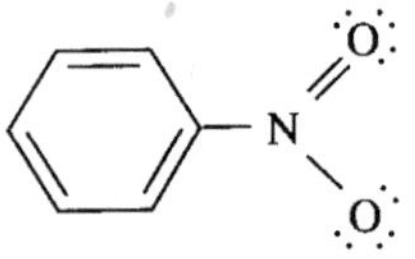 , 4, 1, 4, V, 5

38. 2, 2, 6, VI, zero, 3, 1, 7, -1

39.

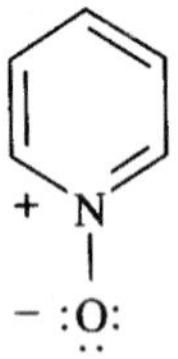

40.

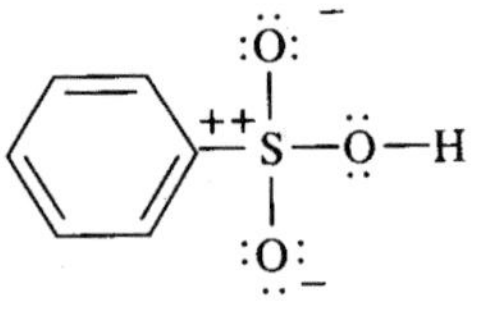

41.

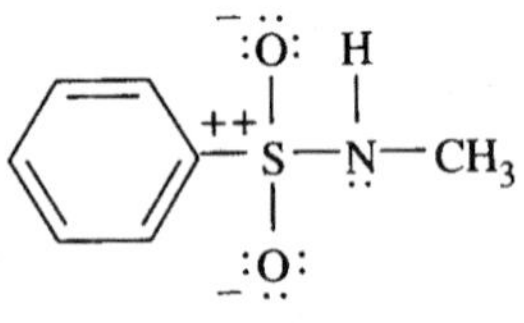

42.

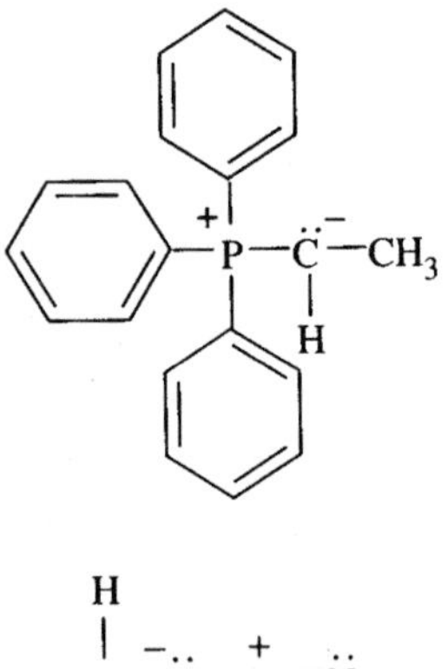

43.

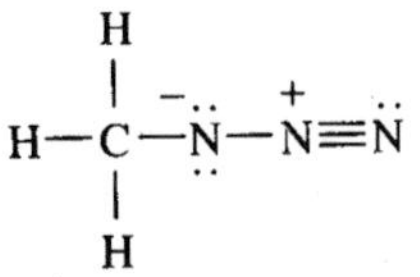

44. $H_2C{=}\overset{+}{N}{=}\ddot{\underset{..}{N}}^-$

45. $C_6H_5{-}\overset{+}{N}{\equiv}\ddot{C}^-$

46. $C_6H_5{-}C{\equiv}\ddot{N}$ , (formal charge is zero on all atoms)

47. $(CH_3)_3\overset{+}{N}{-}\ddot{\underset{..}{O}}{:}^-$

48. $CH_3{-}\overset{+}{\underset{..}{S}}(-\ddot{\underset{..}{O}}{:}^-){-}CH_3$

49. 3, 1, 3, IV, 4, +1, 4, 8, -1

50. $CH_3{-}CH_2{-}CH_2$ , 3, 1, 3, +1, $CH_3{-}CH_2{-}CH_2^+$

51. $CH_3{-}CH(CH_3)$ , 3, 3,+1, $CH_3{-}\overset{+}{C}H(CH_3)$

52. $C_5H_9{-}H$ , 3, 3, +1, $C_5H_8^+{-}H$

53. 3, 1, 5, VI, +1, $CH_3{-}\ddot{O}^+(H){-}H$

54. 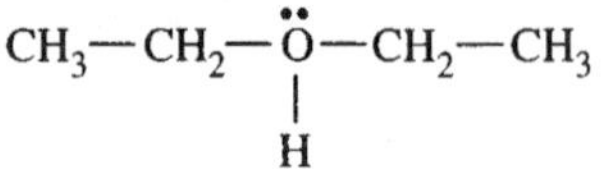 , 3, 1, 5, +1, 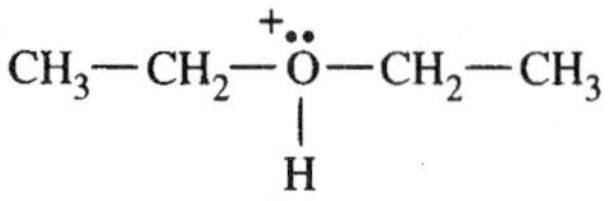

55. 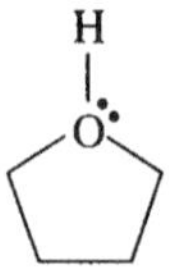 , 3, 1, +1, 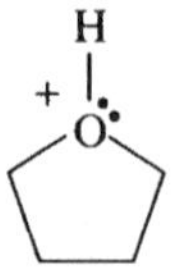

56. 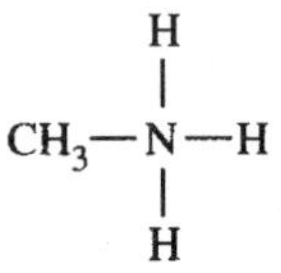 , 4, 1, 4, V, +1, 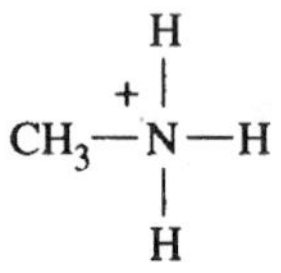

57. 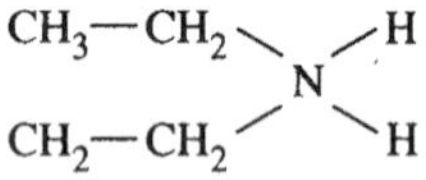 , 4, 4, +1, 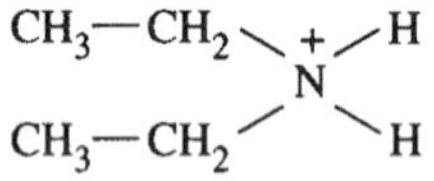

58. 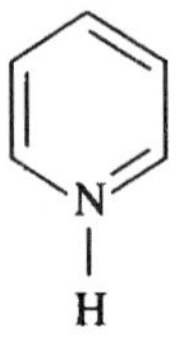 ,4, +1, 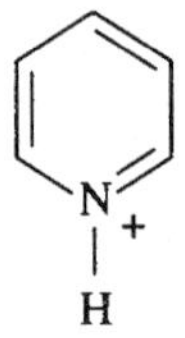

59. 3, 1, 5, IV, +1

60. 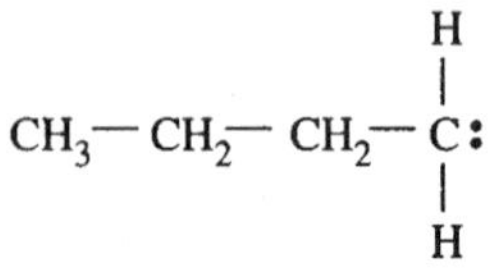 , 3, 1, 5, –1, 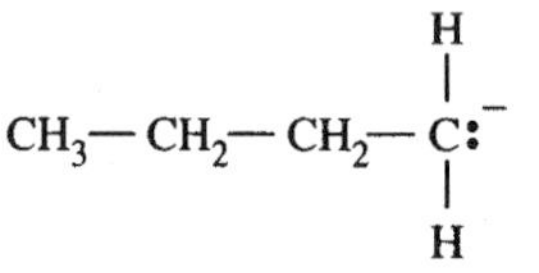

61. 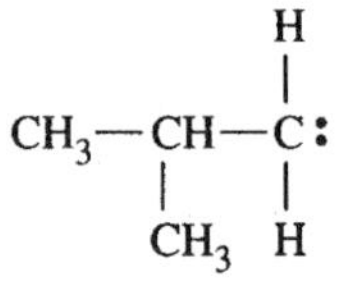 , 3, 1, 5, 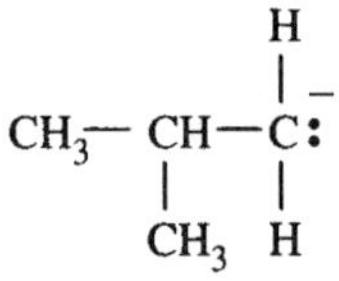

62. 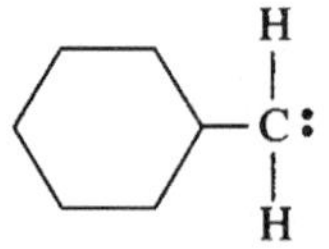 ,3, 1, –1, 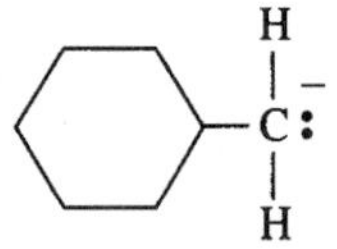

63. 1, 3, 7, VI, –1, 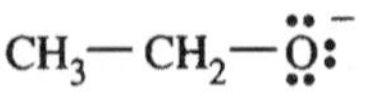

64. 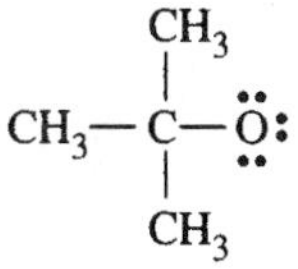 , 1, 3, 7, –1, 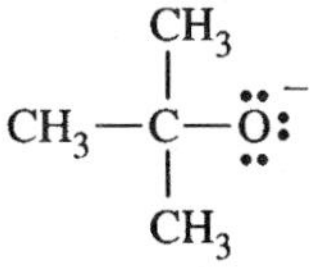

65. 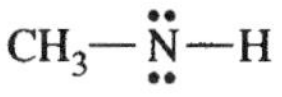 , 2, 2, 6, V, –1, 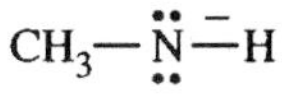

66. 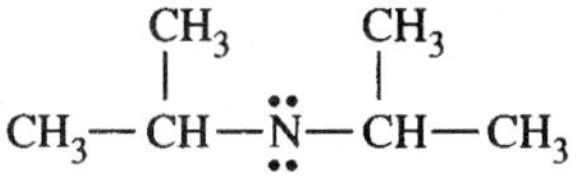 , 6, –1, 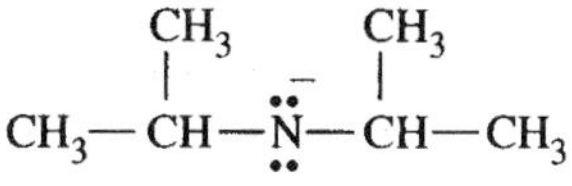

67. 2, 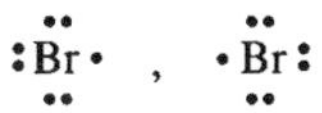

68. 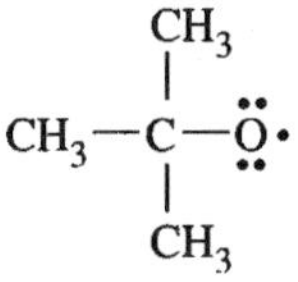 , 1 5, 6, zero

69. 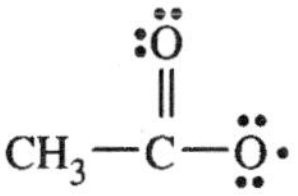 , zero

70. 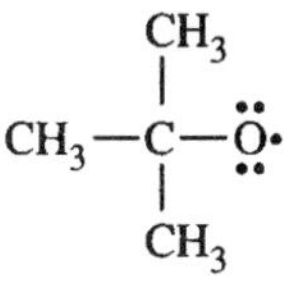

71. 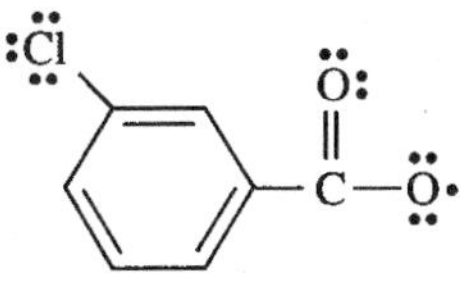 , 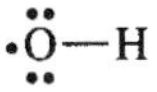

72. 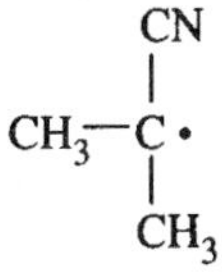

73. 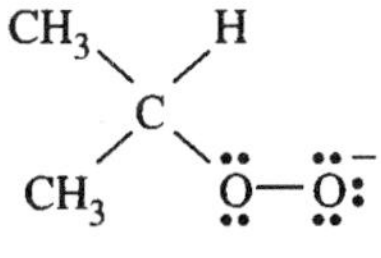

74. $(CH_3)_2CH{-}\ddot{\underset{\cdot\cdot}{O}}\cdot$ , $\cdot\ddot{\underset{\cdot\cdot}{O}}{-}H$

75. $(CH_3)_2\overset{+}{C}H$ , $:\ddot{\underset{\cdot\cdot}{O}}^{-}{-}\ddot{\underset{\cdot\cdot}{O}}{-}H$

# 2 Resonance Structures

Correct Lewis structures are indispensable to understanding the chemical and physical properties of organic molecules. As you will see in Chapter 3 of this text, they are essential to writing the mechanisms of organic reactions. As helpful as they are, however, single Lewis structures do not always describe an organic compound accurately.

By this time you should be acquainted with the resonance method. The resonance method, while still preserving the useful features of the Lewis structure, is a procedure used to describe accurately a molecule for which a single Lewis structure is not sufficient. The resonance method attempts to describe the true structure of a molecule as a hybrid of several Lewis structures. These Lewis structures are called *resonance structures*. It is *essential* to understand that there is only *one* structure for the actual molecule. That structure is not written down but takes its character from the sum of the resonance structures. Resonance structures have no discrete existence of their own but, taken in combination, describe the true structure of the molecule.

The resonance method consists of writing down all the possible resonance structures and making a judgment about which of the resonance structures will make important contributions to the hybrid. In order to do that, one must be able to write resonance structures.

There are several rules to which acceptable resonance structures for a compound must conform. First, the relative positions of all atoms in all resonance structures for a single compound must be the same. Second, all resonance structures for a single compound must have the same number of paired and unpaired electrons. It follows from this rule that the algebraic sum of formal charge on each resonance structure for a single compound must be the same. Third, the important resonance structures will have comparable energies.

> That is, any resonance structure that has an energy substantially higher than the rest must be discarded. The energy of a particular structure depends on several things (charge distribution, the octet rule, **steric strain,** and so on). Judging the relative energies of a series of structures requires an intuitive sense that you develop through experience. This chapter concentrates solely on how to push electrons to generate a series of resonance structures and leaves the development of judgment to you, your professor, and the main text.

While these rules are important, they are not helpful to the student faced with the task of writing resonance structures. This chapter will present a method for writing resonance structures, starting with one Lewis structure and "pushing electrons" to generate new resonance structures. To do that one must find the proper combination of pushable electrons and places to which electrons can be pushed (receptors). In Lewis structures the pushable electron pairs are the unshared electron pairs on a single atom and the pi electron pairs in multiple bonds. Receptors can be (1) atoms with a formal positive charge, (2) atoms that can tolerate a formal negative charge, and (3) atoms which possess pushable electrons themselves.

# Simple Ions

## Cations

One Lewis structure for the allyl cation is

$$\mathrm{H_2\overset{3}{C}{=}\overset{2}{C}H{-}\overset{+}{\overset{1}{C}}H_2}$$

This structure contains a pair of pushable electrons, namely the **pi electrons** in the double bond between $C_2$ and $C_3$. The structure also contains a positively charged carbon atom at $C_1$ that can act as a receptor. A new resonance structure can be generated by pushing the pair of pi electrons to the receptor. This is signified by the curved arrow with two barbs introduced in Chapter 1.

$$\mathrm{H_2\overset{3}{C}{=}\overset{2}{C}H{-}\overset{+}{\overset{1}{C}}H_2} \quad \text{generates} \quad \mathrm{H_2\overset{+}{\overset{3}{C}}{-}\overset{2}{C}H{=}\overset{1}{C}H_2}$$

**(1)** **(2)**

By pushing electrons to the receptor, the positive charge has been neutralized, but a new positive charge has been generated at $C_3$. Structure 1 can be regenerated from (2) by pushing the pi electrons between $C_1$ and $C_2$ toward the positively charged carbon $C_3$. Structures (1) and (2) are equivalent but not identical. In the resonance notation this is written

$$\left[\mathrm{H_2C{=}CH{-}\overset{+}{C}H_2} \longleftrightarrow \mathrm{H_2\overset{+}{C}{-}CH{=}CH_2}\right]$$

Check to be sure that these two resonance structures obey the rules set out in the beginning of this chapter. Also, check the assignment of formal charge.

1. One Lewis structure for the 2-butenyl cation is $\mathrm{CH_3{-}CH{=}CH{-}\overset{+}{C}H_2}$. A new resonance structure can be generated by pushing the pi electrons to the receptor.

$\mathrm{CH_3{-}CH{=}CH{-}\overset{+}{C}H_2}$ generates ____________________

supply arrow

2. A second resonance structure for the 3-cyclopentenyl cation can be generated.

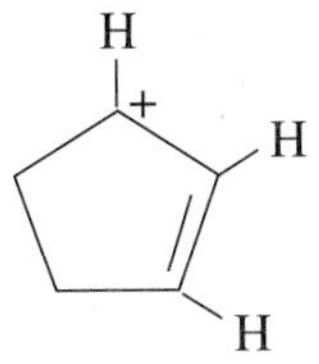

H H H gives

______________________ ______________________

supply arrow

3. One structure for the conjugate acid of acetone is

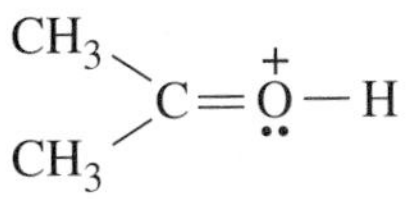

$$(CH_3)_2C{=}\ddot{O}^{+}{-}H$$

The _____ electrons in the carbon–oxygen double bond are pushable electrons, and the _____ atom is a receptor.

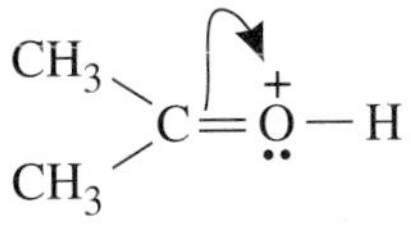

$$(CH_3)_2C{=}\ddot{O}^{+}{-}H$$ generates

______________________

4. Similarly, a resonance structure for the conjugate acid of 2-butanone can be written. Thus,

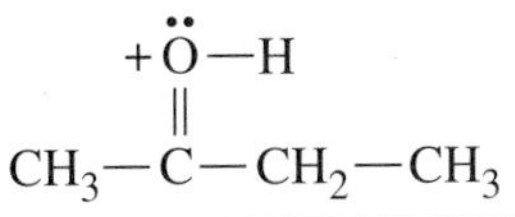

$$CH_3{-}\underset{\|}{C}(=\ddot{O}^{+}{-}H){-}CH_2{-}CH_3$$ generates

______________________ ______________________

supply arrow

It is crucial that these arrows indicate precisely *what* electrons are being pushed *where.* Therefore, after doing these exercises check the answers carefully to see that you have inserted the arrows properly.

5. A second resonance structure for protonated cycloheptanone is produced by

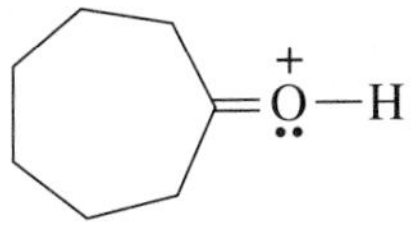

$=\ddot{O}^{+}{-}H$ giving

______________________ ______________________

supply arrow

6. Pairs of unshared electrons can be pushed. One Lewis structure for the methoxymethyl cation is $CH_3-\ddot{\underset{..}{O}}-\overset{+}{C}H_2$. The structure contains a pair of pushable electrons, namely the unshared electrons on the __________ atom. The structure also contains a positively charged __________ atom that can act as a __________. A second resonance structure can be generated by pushing the unshared electrons to the receptor. Thus,

$$CH_3-\ddot{\underset{..}{O}}-\overset{+}{C}H_2 \qquad \text{generates}$$

______________________________ ______________________________

supply arrow

It is not possible to push electrons toward the other carbon, because it is not a receptor.

7. One structure for the acetoxonium ion is

$$CH_3-\overset{+}{C}=\ddot{O}:$$

**(3)**

Clearly, the receptor is the positively charged __________atom. However, there are available two different kinds of pushable electrons, namely, pi electrons in the carbon–oxygen double bond or unshared electrons on the oxygen atom. The proper choice in this case is dictated by the following considerations. In structure (3) the carbon atom possesses a formal positive charge and also has only six electrons in its outer shell (it lacks a stable octet of electrons). If the pi electrons are pushed, namely,

$$CH_3-\overset{+}{C}=\ddot{O}: \longleftrightarrow CH_3-\ddot{C}-\ddot{O}:^{+}$$

**(4)**

structure (4) is obtained in which the oxygen possesses a formal positive charge but, in addition, both the oxygen *and* the carbon atom lack a stable octet. This structure will be very unstable. Structure (4), despite the fact that it can be generated by properly pushing electrons, is not included in the resonance hybrid for the acetoxonium ion. If, on the other hand, the unshared electrons on the oxygen are pushed, a more acceptable structure is obtained, namely,

$$CH_3-\overset{+}{C}=\ddot{O}: \longleftrightarrow$$

__________________

**(5)**

In structure (5) the oxygen atom possesses a formal positive charge, and both the carbon and the oxygen atom have a stable octet. Structure (5) is included in the resonance hybrid.

8. Another ion of this type is

$$CH_3-\underset{\substack{|\\ CH_3}}{CH}-\overset{+}{C}=\ddot{O}:$$

____________________________

supply arrow

which gives the resonance structure

____________________________

*   *   *   *   *

It is important to recognize opportunities to push electrons and generate new resonance structures. It is also important to recognize those structures in which one *cannot* push electrons. In order to be able to generate one or more resonance structure, a Lewis structure must have pushable electrons and a receptor. Furthermore, the receptor must be *next to* the pushable electrons.

9. There are no important resonance structures for the isopropyl cation

$$CH_3-\overset{+}{C}H-CH_3$$

because there are no _________ _________ in the structure. There are no important resonance structures for dimethyl ether

$$CH_3-\ddot{\underset{\cdot\cdot}{O}}-CH_3$$

because, although there are pushable electrons on the oxygen atom, there is no ________. There are no important resonance structures for the 5-pentenyl cation

$$H_2C=CH-CH_2-CH_2-\overset{+}{C}H_2$$

because the pushable electrons and the receptor are separated by two methylene groups. Thus, the electrons have no way to get to the receptor.

## Exercises

In this exercise one Lewis structure is provided. Where possible, push electrons to generate another resonance structure. In some cases no additional structures can be generated.

10. $(CH_3)_2C{=}CH{-}\overset{+}{C}H{-}CH_3 \longleftrightarrow$

11. $(CH_3)_2CH{-}CH{=}\overset{+}{\ddot{O}}{-}H \longleftrightarrow$

12. Cyclohexene ring bearing H on a cationic (+) carbon, with H, H and H substituents and H on the lower ring carbon $\longleftrightarrow$

13. Cyclopentyl$-CH_2{-}\overset{+}{C}H{-}\ddot{O}{-}H \longleftrightarrow$

14. $CH_3{-}\overset{+}{C}(CH_3){-}CH_2{-}\ddot{O}{-}CH_3 \longleftrightarrow$

15. $CH_3{-}C(CH_3)_2{-}\overset{+}{C}H{-}\ddot{O}{-}CH_2{-}CH_3 \longleftrightarrow$

16. Cyclohexyl$-CH_2{-}CH_2{-}\overset{+}{C}{=}\ddot{O}{:} \longleftrightarrow$

17. $\overset{+}{O}{\equiv}C{-}$Cyclopentyl $\longleftrightarrow$

## *Anions*

One Lewis structure for the acetate ion is

This structure contains several pairs of pushable electrons. Since the object of writing resonance structures is to show delocalization of charge, it is reasonable that electrons should be pushed *away* from a center of negative charge. Pushing a pair of unshared electrons, located on the negatively charged oxygen, toward the carbonyl carbon would give

leading to

This, however, gives the intolerable situation of having ten valence electrons around the carboxyl carbon. The situation can be relieved by pushing the pi electrons of the original carbon–oxygen double bond toward the oxygen atom, which can tolerate the negative charge. Thus,

gives

18. The cyclohexane carboxylate anion has a Lewis structure

    Pushing a pair of unshared electrons away from the negatively charged oxygen atom and, at the same time, pushing a pair of pi electrons toward the other oxygen will generate a second resonance structure. Thus,

    ______________ ⟷ ______________

    supply arrows

19. One Lewis structure for the enolate anion of acetaldehyde is

Pushing the pair of unshared electrons on the carbon atom away from the center of negative charge and pushing the pi electrons of the carbon–oxygen double bond to the oxygen atom generates a second resonance structure. Thus,

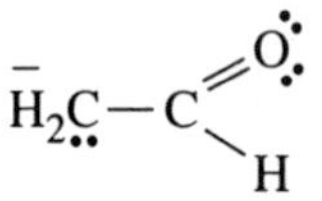

$$\overset{-}{H_2\ddot{C}}-C(=\ddot{O}:)H \qquad \text{gives} \qquad \_\_\_\_\_\_\_\_$$

supply arrows

20. The allyl anion is a resonance hybrid. One of the two equivalent resonance structures is

$$^{-}\ddot{C}H_2-CH=CH_2$$

Delocalization of the unshared electrons on the negatively charged carbon atom and simultaneous delocalization of the pi electrons to the other terminal carbon atom generates another structure.

$$^{-}\ddot{C}H_2-CH=CH_2 \quad \longleftrightarrow \quad \_\_\_\_\_\_\_\_$$

supply arrows

21. The acetonitrile anion provides a slight variation on this theme. Thus, one pair of pi electrons in a triple bond is pushed.

$$^{-}\ddot{C}H_2-C\equiv N: \quad \longleftrightarrow \quad \_\_\_\_\_\_\_\_$$

supply arrows

* * * * *

Here are several errors and misconceptions that appear frequently in attempts to write resonance structures.

1. Always push *electrons,* never push positive charges. Remember that the arrows used in generating resonance structures indicate how the electrons are moving. A structure such as

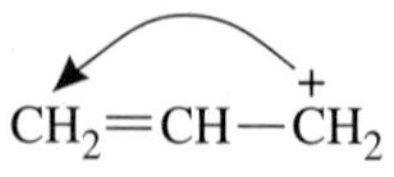

$$CH_2=CH-\overset{+}{C}H_2$$

is incorrect.

2. Always push electrons away from the centers of negative charge and toward centers of positive charge. The resonance method depends on delocalizing charge. Electrons, which have negative charge, must be pushed away from centers of relatively high electron density and toward centers of relatively low electron density.

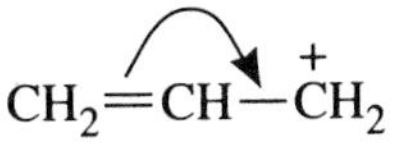

$CH_2{=}CH{-}\overset{+}{C}H_2$
correct

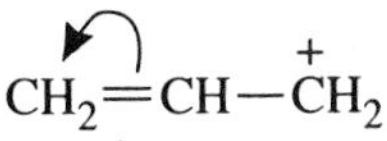

$CH_2{=}CH{-}\overset{+}{C}H_2$
incorrect

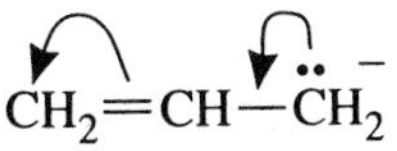

$CH_2{=}CH{-}\ddot{C}H_2^-$
correct

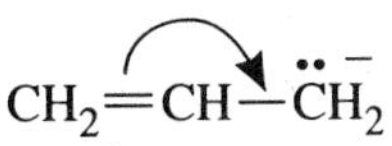

$CH_2{=}CH{-}\ddot{C}H_2^-$
incorrect

## Exercises

In these exercises one Lewis structure is provided. Where possible, push electrons to generate another resonance structure. In other cases there are no additional resonance structures.

22. 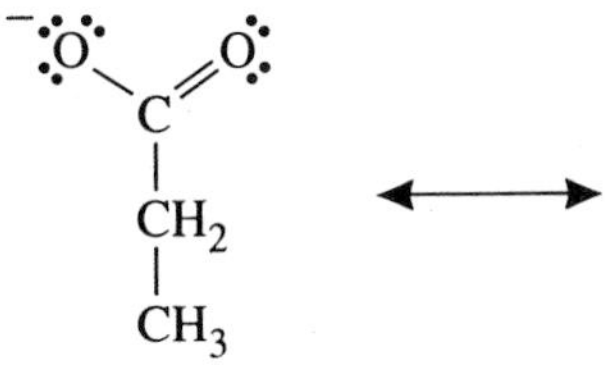 $^-:\ddot{O}{-}C({=}\ddot{O}:){-}CH_2{-}CH_3 \longleftrightarrow$

23. 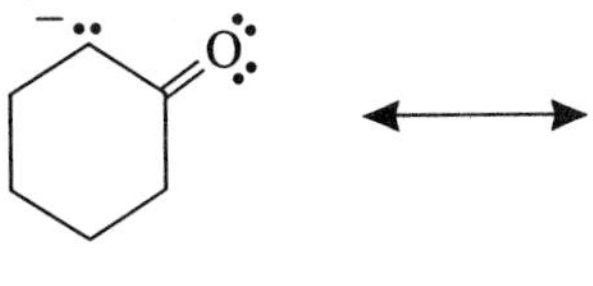 cyclohexanone enolate: ring carbon bearing $^-\ddot{}$ adjacent to $C{=}\ddot{O}:$ $\longleftrightarrow$

24. $:N{\equiv}C{-}\ddot{C}H^-{-}CH_2{-}CH{=}CH_2 \longleftrightarrow$

25. 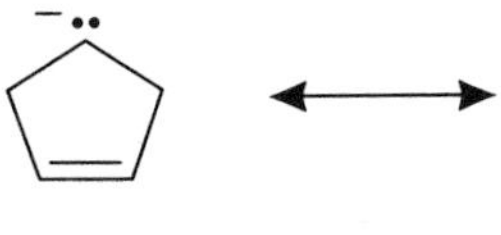 cyclopentene anion ($^-\ddot{}$) $\longleftrightarrow$

26. 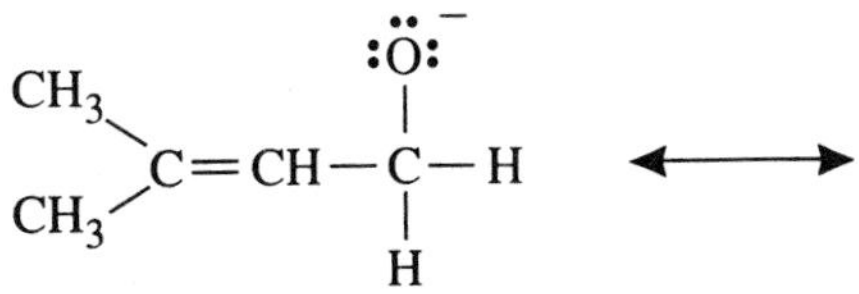 $(CH_3)_2C{=}CH{-}C(:\ddot{O}:^-)(H){-}H \longleftrightarrow$

27. (cyclohexenyl)–$\ddot{C}H_2^-$ ⟷

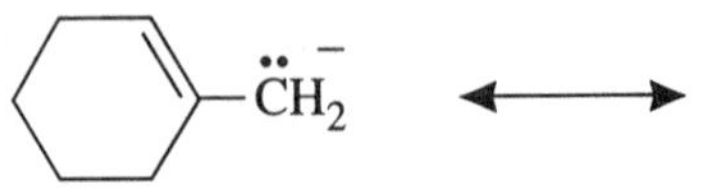

28. (cyclopentyl)–$\ddot{C}H^-$–(cyclopentyl) ⟷

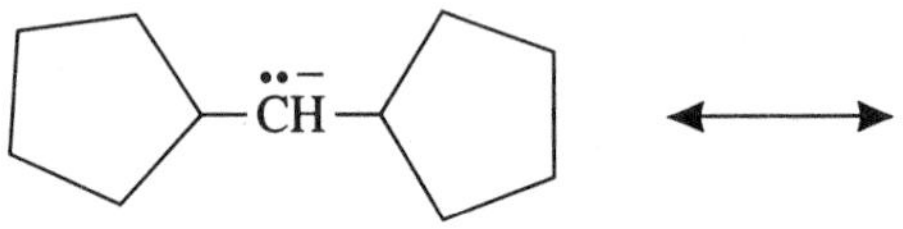

## Free Radicals

Some free radicals are stabilized by electron delocalization. The resonance method depicts this well. The formalism is a bit different from that used for ions and you must pay close attention at first.

Electrons are pushed as pairs (arrows with two barbs) when writing resonance structures for molecules or ions. For free radicals, electrons are pushed individually (arrows with one barb) and some electrons become unpaired as others become paired.

The allyl radical has an unpaired electron and a pair of pi electrons located next to each other. The generation of an equivalent, but not identical, resonance structure is shown as the pi-pair becoming unpaired (a homolytic cleavage of the **pi bond**) and one of those electrons becoming paired with the erstwhile unpaired electron.

$$CH_2{=}CH{-}\dot{C}H_2 \longleftrightarrow \dot{C}H_2{-}CH{=}CH_2$$

The notation is cluttered with more arrows since each electron has its own arrow.

29. Draw the second resonance structure.

$$\underset{4}{CH_3}{-}\underset{3}{CH}{=}\underset{2}{CH}{-}\underset{1}{\dot{C}H_2} \longleftrightarrow \underline{\qquad\qquad}$$

Note that no delocalization occurs at C4.

30. Draw another resonance structure.

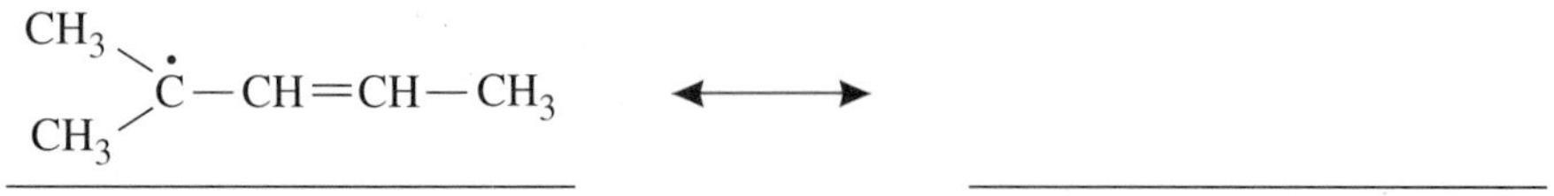

$$(CH_3)_2\dot{C}{-}CH{=}CH{-}CH_3 \longleftrightarrow \underline{\qquad\qquad}$$

supply arrows

31. Resonance occurs in cyclic radicals.

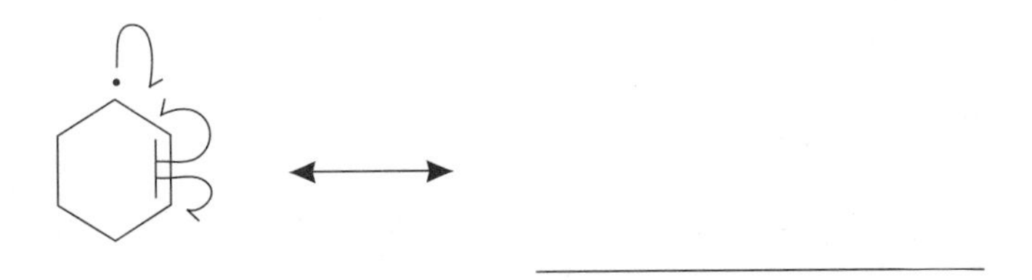

32. Again.

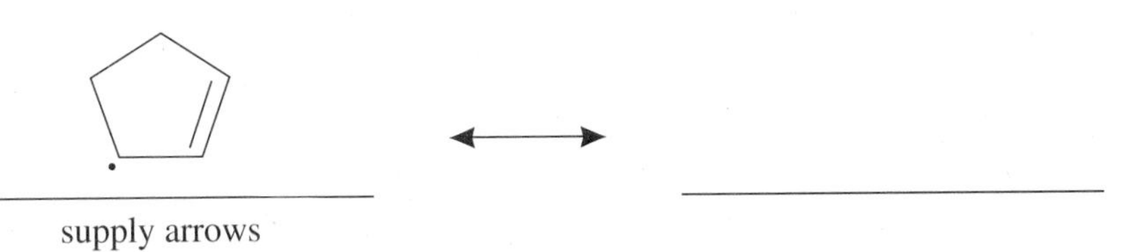

supply arrows

33. Electrons can be delocalized from any multiple bond next to a radical center.

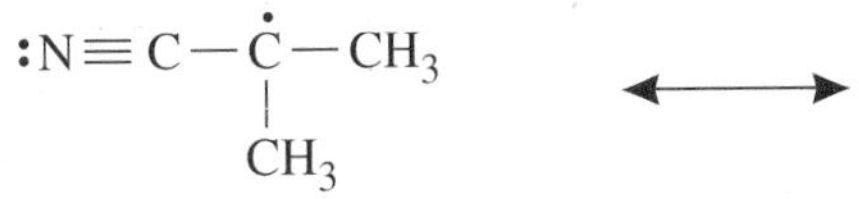

$:N\equiv C-\dot{C}(CH_3)-CH_3$ ⟷ ______

supply arrows

34. Again.

$CH_3(H)\dot{C}-C(=\ddot{O}:)CH_3$ ⟷ ______

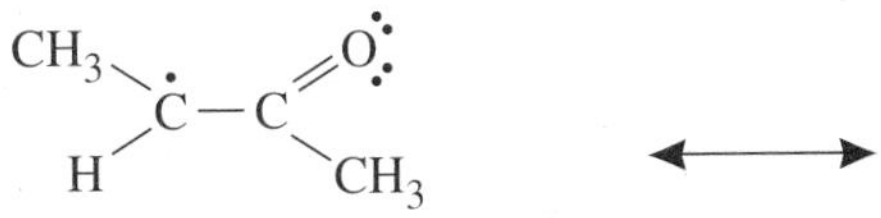

supply arrows

## Exercises

One Lewis structure is provided along with the number of possible structures (including the one provided). Draw the other structures.

35. (3)

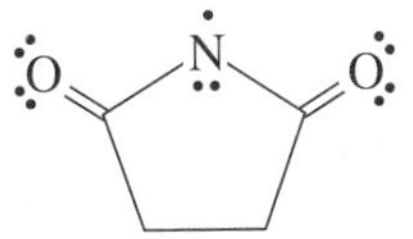

36. (3)

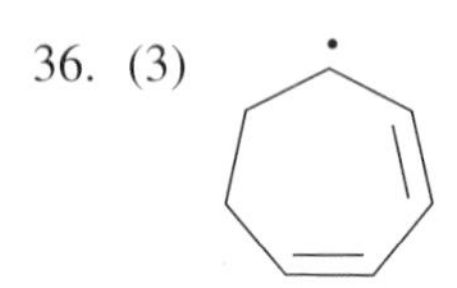

## Benzene and Benzenoid Aromatic Compounds

The best known example of a resonance hybrid is benzene. Benzene has a Lewis structure

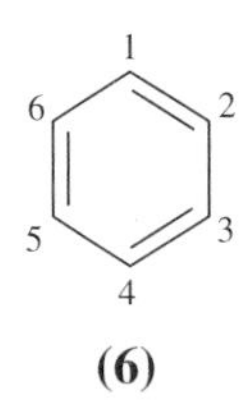


**(6)**

This structure contains three pairs of pushable electrons. It is expedient, however, to consider only one pair, say the pair of pi electrons between C1 and C2, as the pushable electrons and the others as receptors.

37. Drawing an arrow pushing the C1–C2 pi electrons toward C3 would result in the unacceptable situation of having ten electrons around C3. The situation is relieved by pushing the C3–C4 pi electrons toward C5 giving

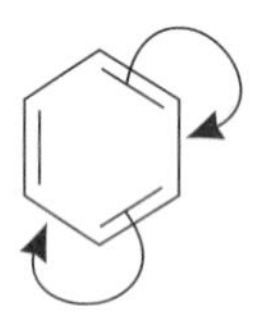

This presents C5 with the same dilemma that C3 had a moment ago. By pushing the C5–C6 pi electrons toward C1, the dilemma is solved since C1 is missing the pair of pushable pi electrons that were used to begin this process. Thus,

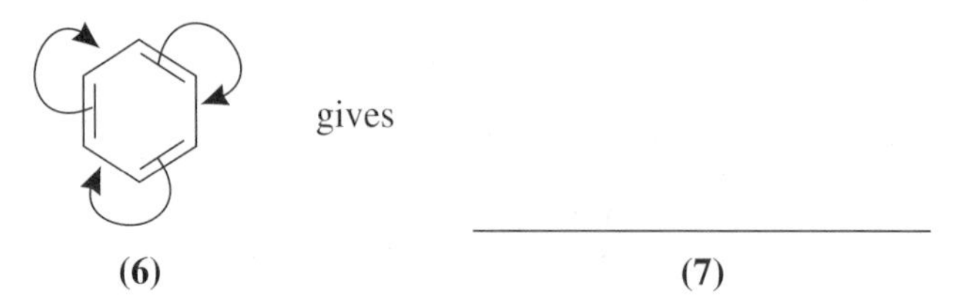


**(6)** **(7)**

Structures (6) and (7) are referred to as Kekulé structures. Whenever a benzene ring appears in a compound, the two Kekulé structures should be the first two resonance structures written.

August Kekulé (1829–96) proposed and proved this structure for benzene in 1865. He recognized at the time that a single structure did not account for certain properties of benzene and proposed, not too convincingly, a dynamic equilibrium between the two equivalent but not identical structures. The resonance theory (1935) resolved this problem.

Several shorthand notations that imply the two Kekulé structures are used by organic chemists. They are

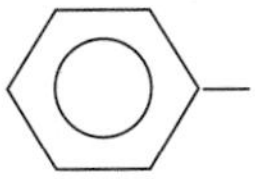

, $C_6H_5$— , and Ph —

Each of these can be used to indicate the phenyl group. Thus, toluene can be written

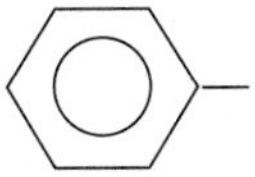

—$CH_3$ , $C_6H_5$—$CH_3$ , Ph—$CH_3$ ,

or

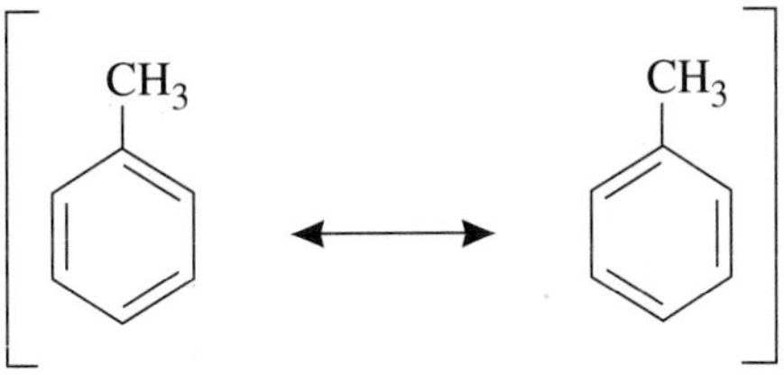

$CH_3$ ⟷ $CH_3$

38. A second resonance structure for *ortho*-xylene can be generated.

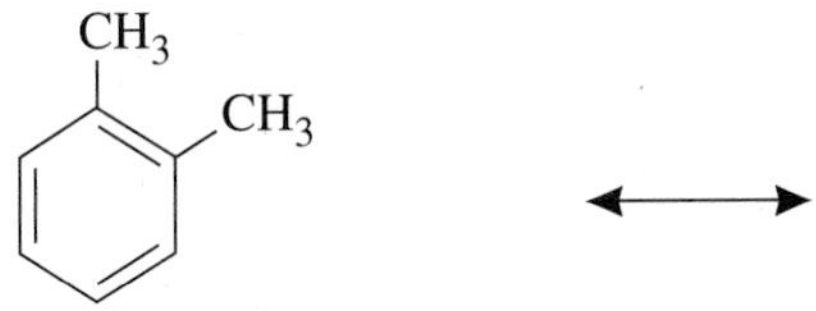

$CH_3$ $CH_3$ ⟷

supply arrows

39. When more than one benzene ring is present in the same molecule, the number of possible resonance structures is increased. In diphenylmethanol, a second resonance structure can be generated by pushing electrons in the ring on the right and leaving the ring on the left undisturbed.

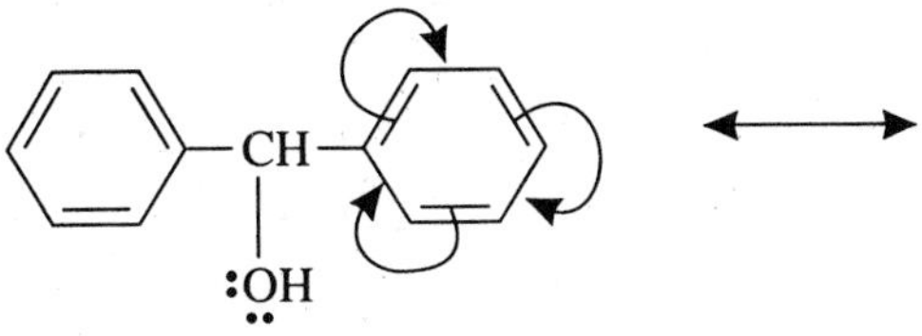

CH :OH ⟷

Conversely, pushing electrons in the ring on the left and leaving the right undisturbed gives another structure.

CH

:OH

supply arrows

Finally, pushing electrons on both rings gives a fourth resonance structure.

CH

:OH

supply arrows

The complete resonance notation for diphenylmethanol is

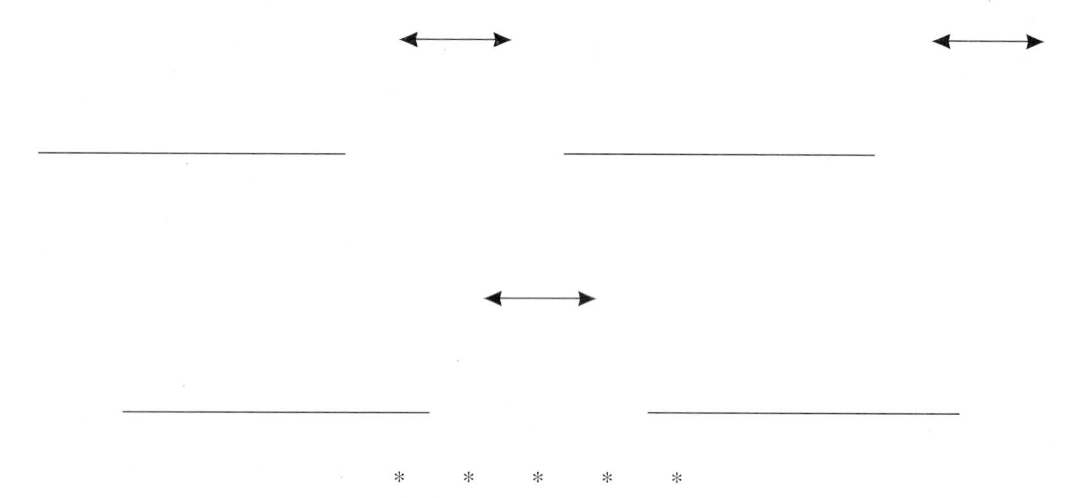

* * * * *

"Fused" aromatic compounds contain benzene rings that have two carbon atoms in common. Writing resonance structures for fused aromatic compounds is done in a way similar to that described for diphenylmethanol. That is, electrons in one ring are pushed and the rest of the molecule is left undisturbed.

40. One Lewis structure for naphthalene is

Pushing electrons in the right ring and leaving the left alone gives

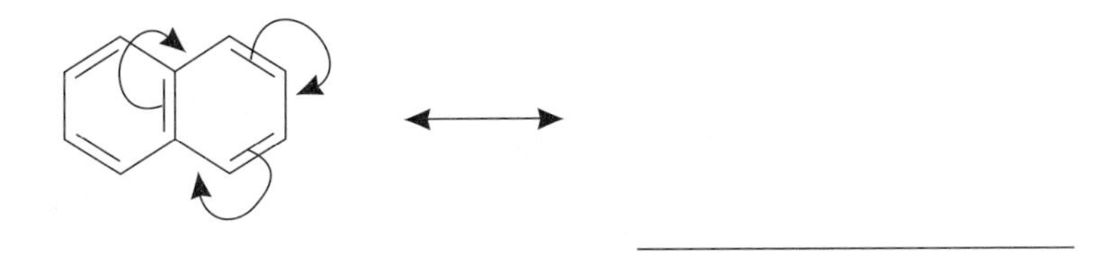

Pushing electrons in the left ring and leaving the right alone gives

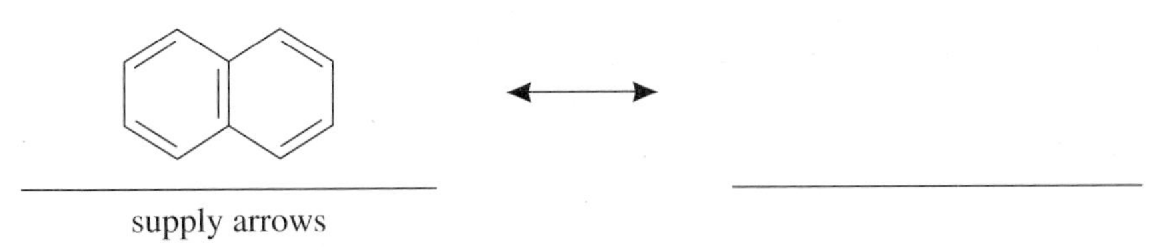

supply arrows

Note that one cannot push electrons in adjacent fused rings at the same time because one pair of pi electrons is held in common between the two.

41. Phenanthrene contains three fused rings and has five acceptable resonance structures. They can be generated as follows.

(a) Push electrons in the left ring.

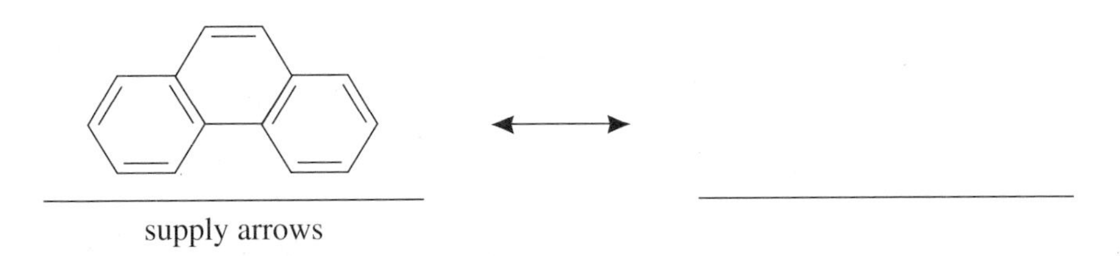

supply arrows

(b) Push electrons into the right ring.

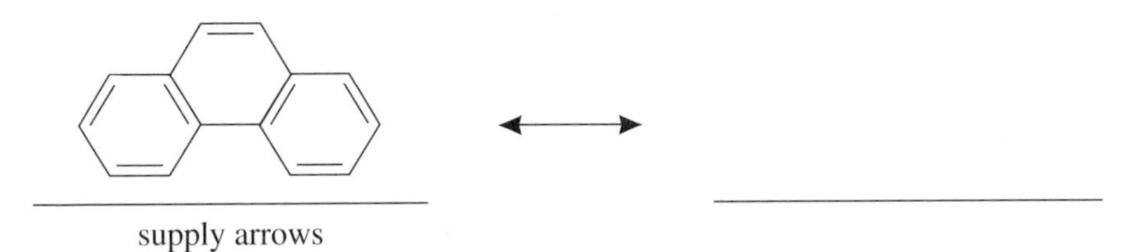

supply arrows

(c) Push electrons in the right and left rings.

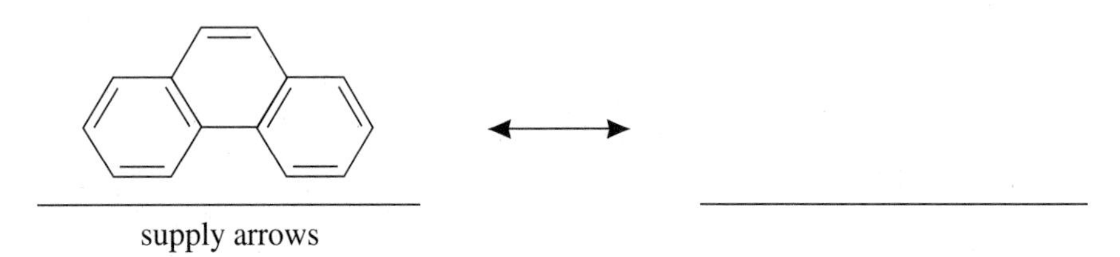

supply arrows

(d) Push electrons in the central ring.

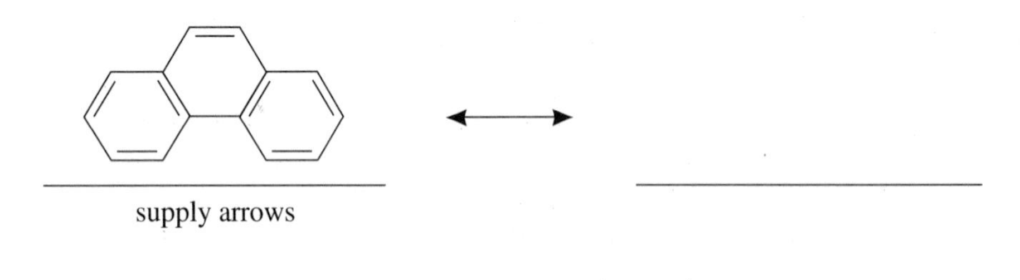

supply arrows

* * * * *

Writing resonance structures for fused aromatic compounds is particularly rich in opportunities for incorrect structures. After all resonance structures for fused aromatic compounds are written down, check for and remove the following.

1. Any structures with ten electrons around carbon, e.g.,

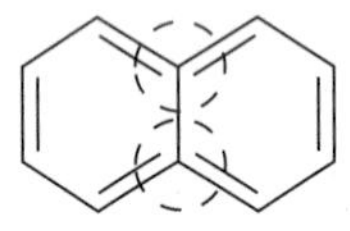

2. Any structures that do not have a continuous system of alternating single and double bonds, e.g.,

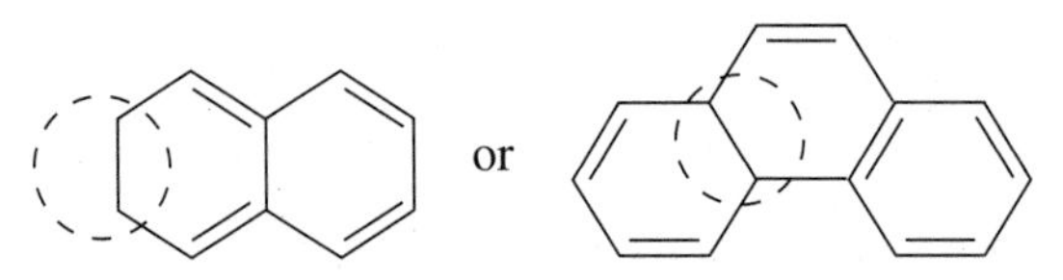


3. Any structures that are identical with others in the set, e.g.,

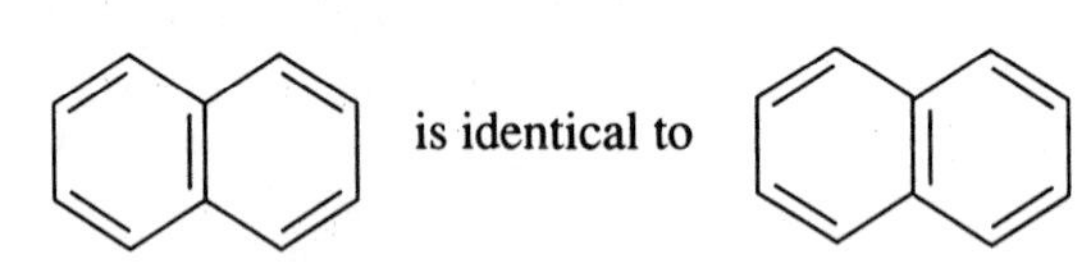

## Exercises

In these exercises, one Lewis structure of an aromatic compound is given along with the total number (including the one provided) of possible structures. Write the other structures.

42. Ethylbenzene (2) $CH_2—CH_3$

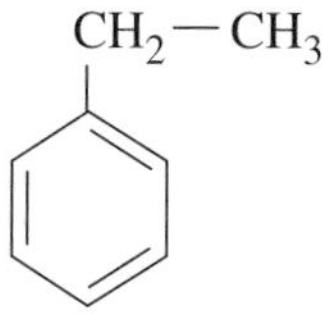

43. p-xylene (2) $CH_3$ $CH_3$

44. Anthracene (4)

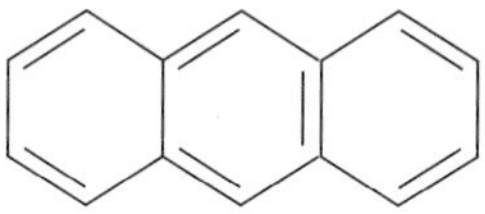

# *More Complicated Ions*

45. Very often more than two resonance structures can be written for an ion. One Lewis structure for the phenoxide ion is

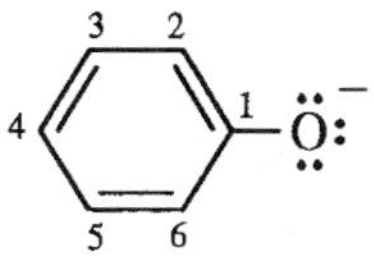

**(8)**

The structure contains a pair of pushable electrons, namely, the unshared electrons on the negatively charged oxygen atom. The structure also contains a receptor, C1, from which a pair of pi electrons can be pushed to C2. This is shown as

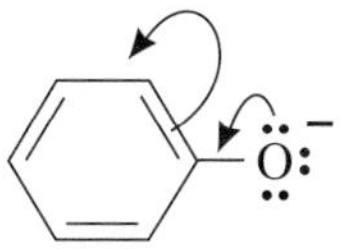

which generates

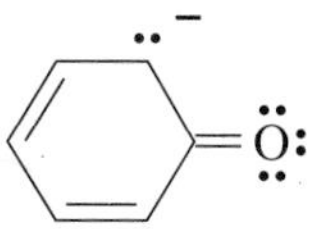

**(9)**

Structure (9) in its turn can be used to generate another resonance structure by pushing the unshared electrons at C2 toward C3 as the C3—C4 pi electrons are pushed to C4.

generates

supply arrows (10)

A final structure, equivalent to but not identical to (9) can be arrived at by pushing the unshared pair at C4 toward C5 and moving the C5—C6 pi electrons to C6.

generates

supply arrows (11)

Try to complete the following without referring to the structures you have just written. The complete resonance method designation for phenoxide ion, including both Kekulé forms of the first structure, is

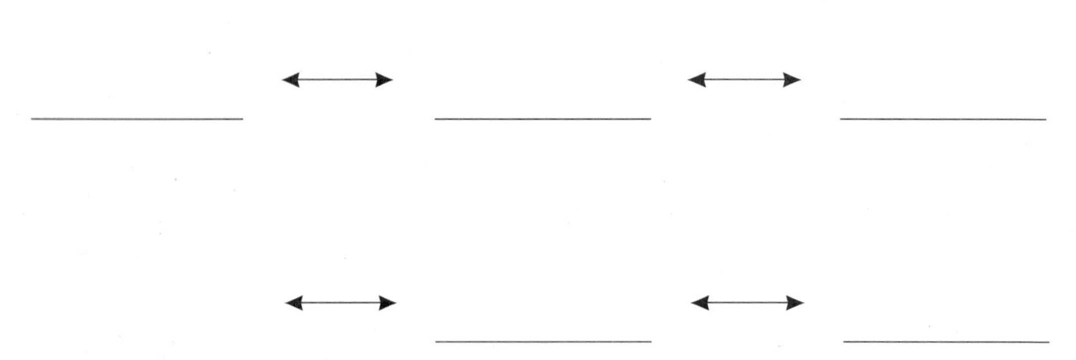

46. One Lewis structure for the cyclopentadienide anion is

(12)

Four additional resonance structures can be written. The pair of unshared electrons at C1 is pushed toward C2, and the pair of pi electrons between C2 and C3 is pushed toward C3.

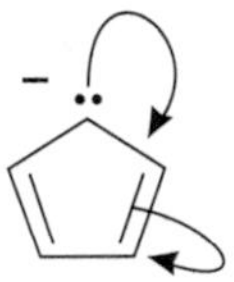

(12) ⟷ ______ (13)

In structure (13) there is a pair of electrons at C3 and double bonds at C4—C5 and C1—C2. The unshared pair at C3 is pushed toward C4 and the pair of pi electrons at C4—C5 is pushed toward C5.

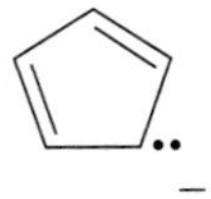

______ ⟷ ______

supply arrows

(13) (14)

This process can be repeated twice more to generate equivalent, but not identical, structures.

______ ⟷ ______

(14) (15)

______ ⟷ ______

(15) (16)

A similar operation on (16) regenerates (12).

47. There are five acceptable resonance structures for the benzyl cation.

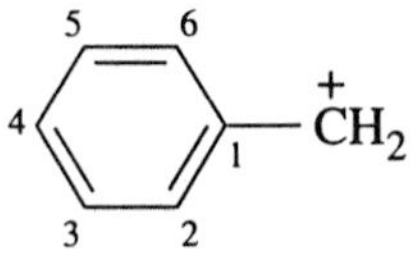

5 6 4 1 3 2 $\overset{+}{C}H_2$

**(17)**

First, the other Kekulé structure.

________________________

**(18)**

Then, a pair of pi electrons from the ring can be pushed toward the positively charged carbon atom.

$\overset{+}{C}H_2$ ⟷ ________________________

**(17)** **(19)**

Structure (19) becomes (20) by pushing the C3–C4 pi electrons.

________________________ ⟷ ________________________

**(19)** **(20)**

A similar operation on (20) generates (21), which is equivalent, but not identical, to (19).

________________________ ⟷ ________________________

**(20)** **(21)**

48. As long as pushable electrons and receptors are available, one can continue to write new

resonance structures. For example, in the bromination of aniline the intermediate, (22), appears.

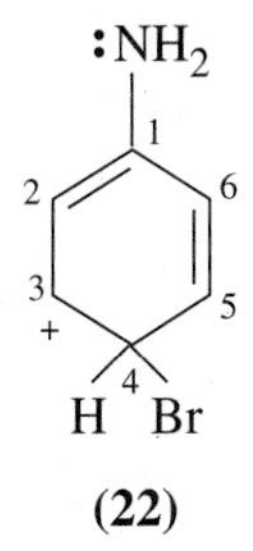


(22)

The pair of pi electrons between C1 and C2 can be pushed toward the positively charged C3, which is a receptor.

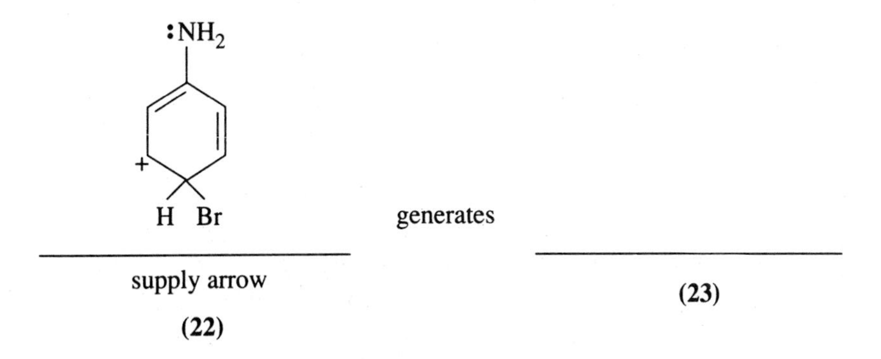


In this new structure the same opportunity presents itself, i.e., to push the pair of pi electrons between C5 and C6 to the positively charged carbon at C1.

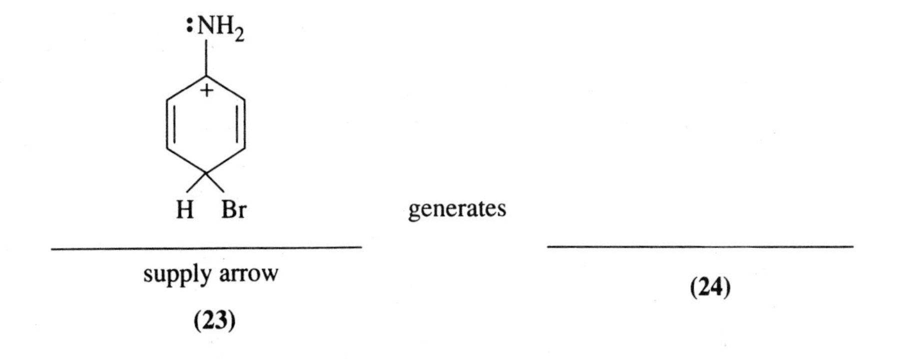

In structure (23) there is another pair of pushable electrons next to the receptor, namely, the unshared pair on the nitrogen atom.

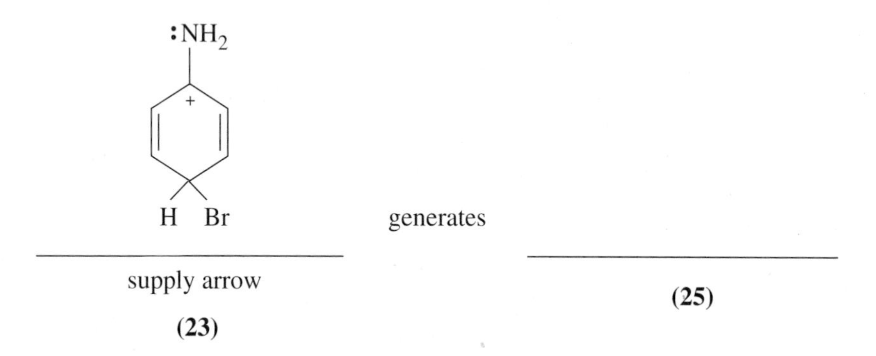


(23) (25)

Don't refer back to complete the following. The complete resonance method notation for this intermediate is

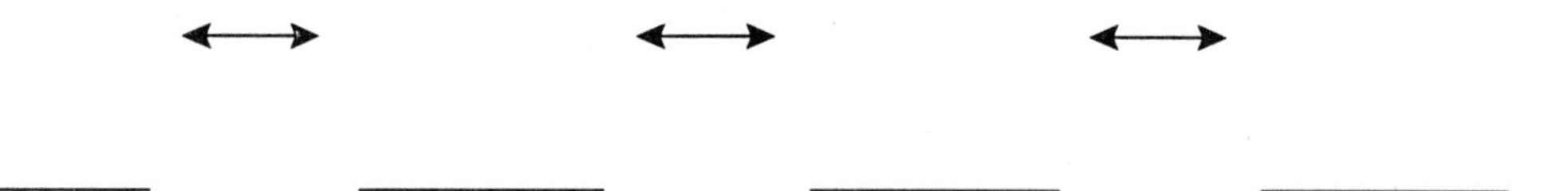

49. The intermediate, (26), appears in the bromination of styrene.

H, Br, 1, 2, 3, 4, 5, 6, +, $-CH{=}CH_2$

(26)

The pair of electrons between C1 and C2 can be pushed toward C3.

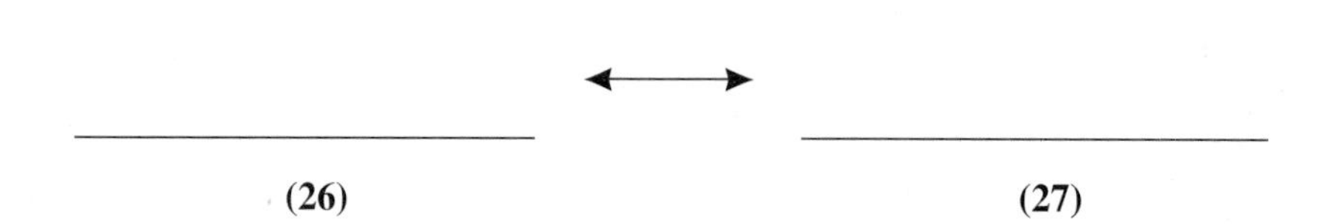

(26) (27)

In (27) the same opportunity presents itself.

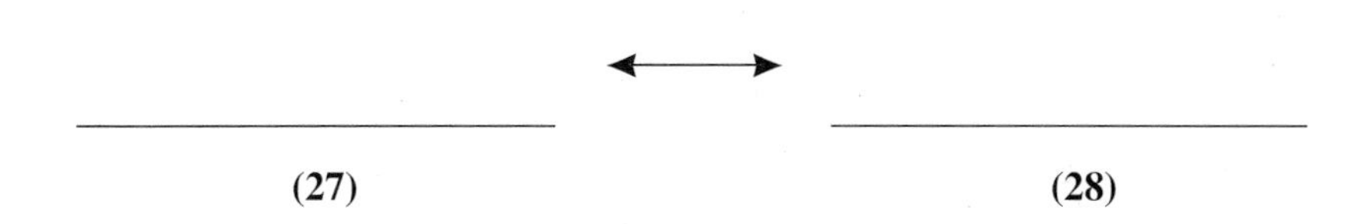


In (27) there is another pair of pushable electrons, namely, the pair of pi electrons in the double bond outside the benzene ring.

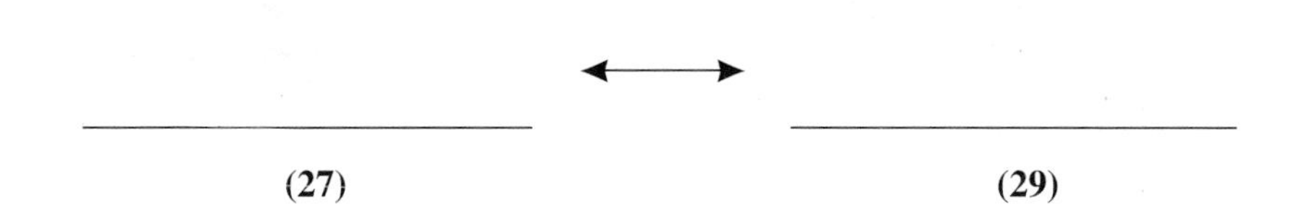


## Exercises

One Lewis structure is provided along with the number of possible structures (including the one provided). Write the other structures.

50. (3)

51. (3)

52. (7)

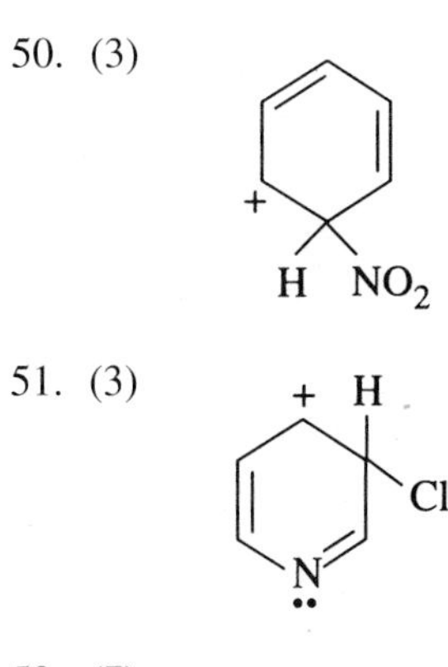

53. (3)

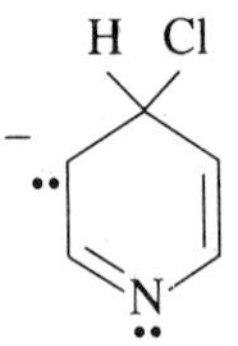

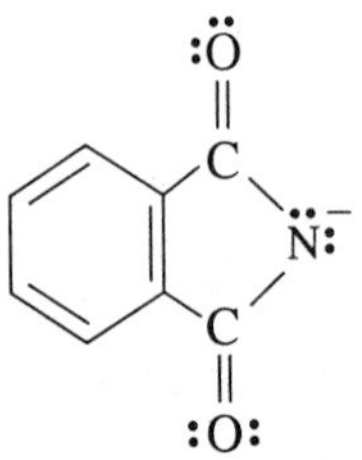

55. (4)

54. (3)

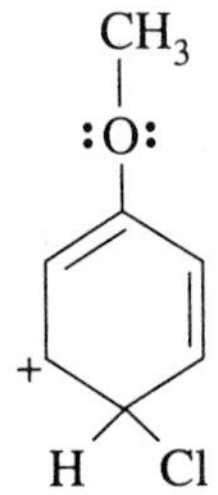

## Molecules Having Resonance Structures with Charge Separation

So far in this chapter all the resonance structures for a given molecule, ion, or radical have possessed the same formal charge distribution. These resonance structures are significant because the energies of the various structures are similar and, thus, each structure makes an important contribution to the hybrid.

In many cases, however, molecules for which there is a Lewis structure with no formal charge separation also have resonance structures with charge separation. These structures make smaller, albeit finite, contributions to the hybrid.

The Lewis structure for formaldehyde is

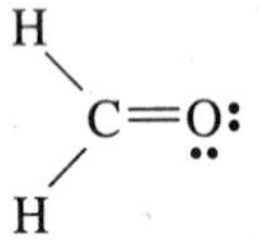

Of the two atoms, carbon and oxygen, that are sharing the pi electrons of the carbonyl bond, oxygen is the more electronegative. Therefore, it can act as a receptor. Pushing the pi electrons to the oxygen atom generates a new resonance structure.

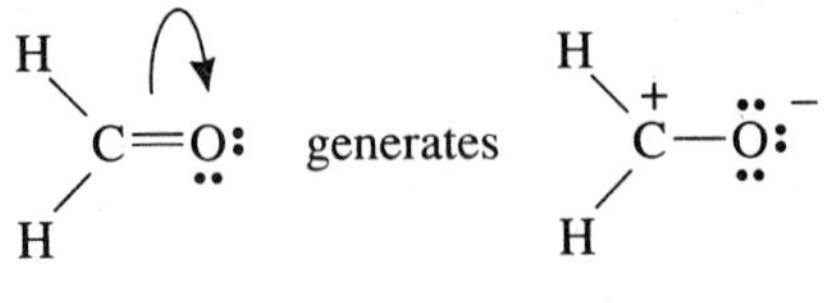

generates

The introduction of charge separation into a resonance structure usually increases the energy of that structure relative to one in which no separation of charge appears. The uncharged structure makes a substantially greater contribution to the hybrid. However, proposing that the charged structure makes a smaller but finite contribution emphasizes the **electrophilic** nature of a carbonyl carbon atom and the **nucleophilic** nature of the carbonyl oxygen atom. An alternative structure in which the pi electrons are pushed toward the carbon atom, i.e.,

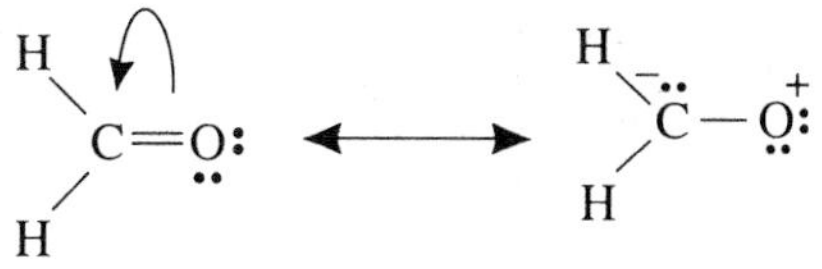

$H_2C{=}\ddot{O}{:} \longleftrightarrow H_2\overset{-}{\ddot{C}}{-}\overset{+}{\ddot{O}}{:}$

No good!

is not considered because it makes no sense to push electrons away from the more electronegative atom.

Neither is it acceptable to propose a diradical resonance structure since that violates the principle that all resonance structures for a single molecule must contain the same number of paired and unpaired electrons.

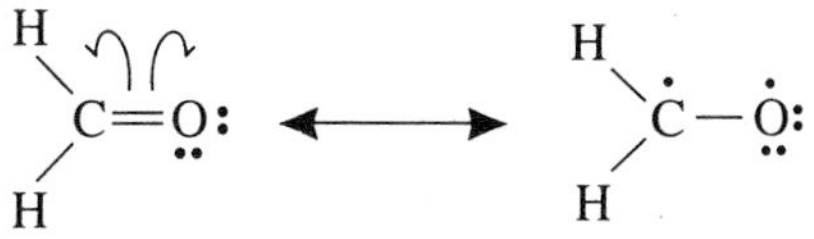

$H_2C{=}\ddot{O}{:} \longleftrightarrow H_2\dot{C}{-}\dot{\ddot{O}}{:}$

No good!

56. The Lewis structure for acrolein (propenal) is

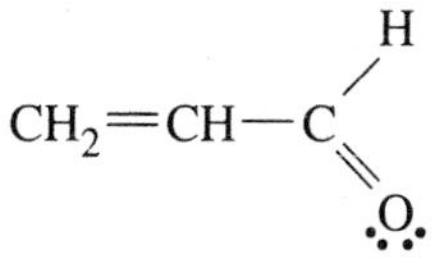

$CH_2{=}CH{-}C(H){=}\ddot{O}{:}$

Pushing the pi electrons of the carbonyl bond toward the oxygen atom will generate a second resonance structure.

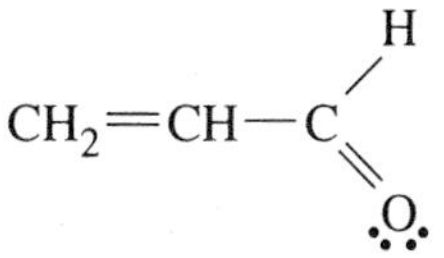

$CH_2{=}CH{-}C(H){=}\ddot{O}{:}$ generates ______________________

______________________
supply arrow

The second structure has a pair of pushable electrons (namely, the pi electrons of the carbon–carbon double bond) next to a receptor, the carbonyl carbon atom, which possesses a formal positive charge.

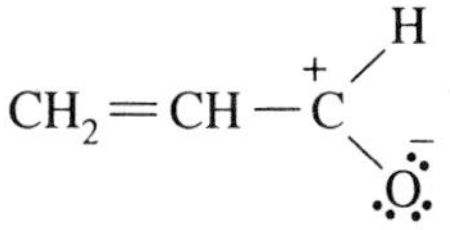

$CH_2{=}CH{-}\overset{+}{C}(H){-}\overset{-}{\ddot{O}}{:}$ gives ______________________

______________________
supply arrow

Don't refer back to complete the following. The resonance method notation for acrolein is

57. The Lewis structure for N-methylacetamide is shown. Pushing the pi electrons of the carbonyl bond toward the oxygen atom will generate a new resonance structure.

$$CH_3-\overset{\overset{\ddot{O}:}{\|}}{C}-\underset{\cdot\cdot}{N}\begin{matrix}CH_3\\H\end{matrix}$$

______________________ gives ______________________

supply arrow

The unshared electrons on the nitrogen atom in the second structure can be pushed to the positively charged carbonyl carbon and

$$CH_3-\overset{\overset{:\ddot{O}:^-}{|}}{\underset{+}{C}}-\underset{\cdot\cdot}{N}\begin{matrix}CH_3\\H\end{matrix}$$

______________________ gives ______________________

supply arrow

Don't refer back to complete the following. The resonance method notation for N-methylacetamide is

58. Once again the presence of an aromatic ring greatly enhances the opportunities to push electrons and generate new resonance structures. The nitro group in nitrobenzene is a powerful **electron attractor**. A pair of pi electrons from the ring can be pushed toward the nitrogen atom as the pair

of pi electrons in the N–O double bond are pushed toward the oxygen atom. That is,

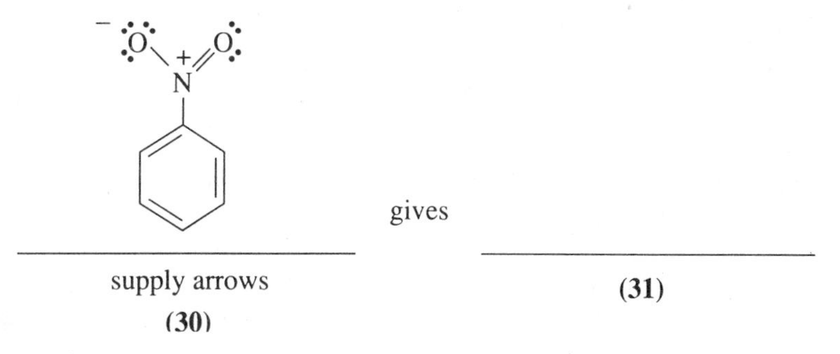

gives ______________

supply arrows
**(30)**

**(31)**

In (31) a pair of pi electrons in the ring can be pushed toward the positive charge.

generates ______________

supply arrow
**(31)**

**(32)**

The process can be repeated

generating ______________

supply arrow
**(32)**

**(33)**

Fill in the complete resonance method notation for nitrobenzene. [Watch out for three other structures that can be generated from (30).]

59. Hypervalence (p. 12) results in an interesting resonance structure with separation of charge. A phosphonium ylide is an intermediate in the Wittig synthesis of alkenes. The phosphorous atom in the ylide is hypervalent. By pushing pi electrons away from the phosphorous one generates resonance structure with charge separation but with every atom obeying the octet rule.

$(Ph)_3P{=}CH_2$ ⟷ ________

phosphonium ylide

## Exercises

One Lewis structure is provided along with the total number of resonance structures (including the one provided). Write the others.

60. 2-cyclohexenone (3)

61. Ethyl acetate (3)

$CH_3{-}C({=}O){-}O{-}CH_2{-}CH_3$

62. Propionic acid (3)

$H{-}O{-}C({=}O){-}CH_2{-}CH_3$

63. Benzonitrile (7)

$C_6H_5{-}C{\equiv}N$:

64. An ylide (2)

$(Ph)_3P{=}C(H)(CH_3)$

# *Answers*

---

## *Chapter 2*

1. 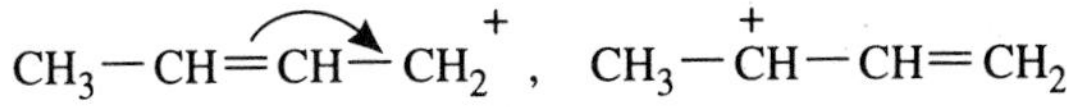 $CH_3-CH=CH-\overset{+}{C}H_2$ , $CH_3-\overset{+}{C}H-CH=CH_2$

2. H H H H H H + +

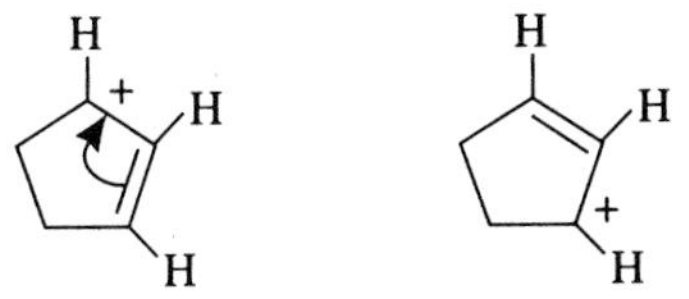

3. pi, oxygen, $(CH_3)_2\overset{+}{C}-\ddot{O}-H$

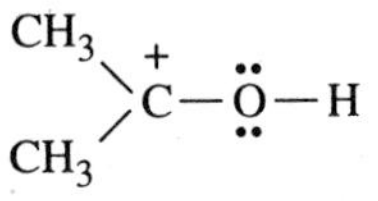

4. 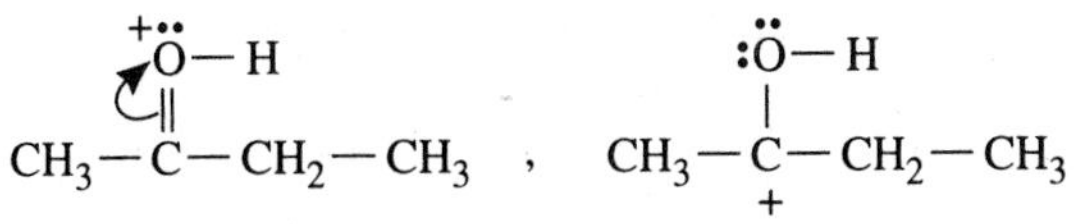 $CH_3-\underset{\|}{C}(\overset{+}{\ddot{O}}-H)-CH_2-CH_3$ , $CH_3-\underset{+}{C}(\ddot{O}-H)-CH_2-CH_3$

5. 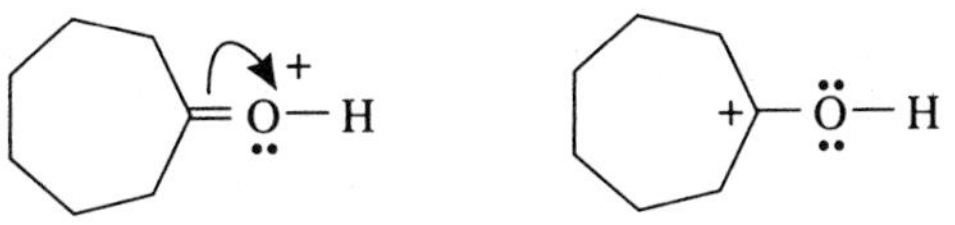 $=\overset{+}{O}-H$ , $+-\ddot{O}-H$

6. oxygen, carbon, receptor,

$CH_3-\ddot{O}-\overset{+}{C}H_2$ , $CH_3-\overset{+}{O}=CH_2$

7. carbon, $CH_3-C\equiv\overset{+}{O}$

8. 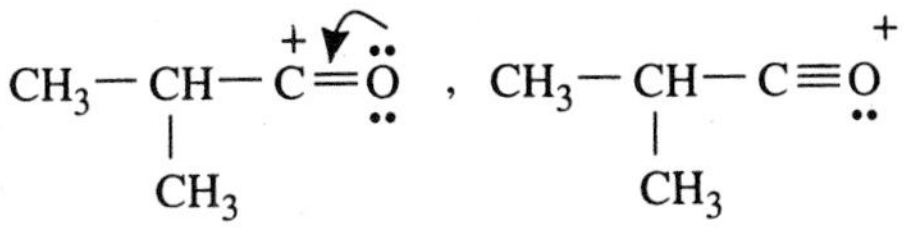 $CH_3-CH(CH_3)-\overset{+}{C}=\ddot{O}$ , $CH_3-CH(CH_3)-C\equiv\overset{+}{O}$

9. pushable electrons, receptor

10. $(CH_3)_2C{=}CH{-}\overset{+}{C}H{-}CH_3 \longleftrightarrow (CH_3)_2\overset{+}{C}{-}CH{=}CH{-}CH_3$

11. $(CH_3)_2CH{-}CH{=}\overset{+}{\ddot{O}}{-}H \longleftrightarrow (CH_3)_2CH{-}\overset{+}{C}H{-}\ddot{\underset{\cdot\cdot}{O}}{-}H$

12. no additional structures

13. $C_5H_9{-}CH_2{-}\overset{+}{C}H{-}\ddot{\underset{\cdot\cdot}{O}}{-}H \longleftrightarrow C_5H_9{-}CH_2{-}CH{=}\overset{+}{\underset{\cdot\cdot}{O}}{-}H$

14. no additional structures

15. $CH_3{-}C(CH_3)_2{-}\overset{+}{C}H{-}\ddot{\underset{\cdot\cdot}{O}}{-}CH_2{-}CH_3 \longleftrightarrow CH_3{-}C(CH_3)_2{-}CH{=}\overset{+}{\underset{\cdot\cdot}{O}}{-}CH_2{-}CH_3$

16. $C_6H_{11}{-}CH_2{-}CH_2{-}\overset{+}{C}{=}\ddot{\underset{\cdot\cdot}{O}} \longleftrightarrow C_6H_{11}{-}CH_2{-}CH_2{-}C{\equiv}\overset{+}{\underset{\cdot\cdot}{O}}$

17. $\overset{+}{\underset{\cdot\cdot}{O}}{\equiv}C{-}C_5H_9 \longleftrightarrow \ddot{\underset{\cdot\cdot}{O}}{=}\overset{+}{C}{-}C_5H_9$

18. $^{-}{:}\ddot{\underset{\cdot\cdot}{O}}{-}C({=}\ddot{O}{:}){-}C_6H_{11} \longleftrightarrow {:}\ddot{O}{=}C(\ddot{\underset{\cdot\cdot}{O}}{:}^{-}){-}C_6H_{11}$

19. $H_2\overset{-}{\underset{\cdot\cdot}{C}}{-}C(H){=}\ddot{O}{:} \longleftrightarrow H_2C{=}C(H){-}\ddot{\underset{\cdot\cdot}{O}}{:}^{-}$

20. $^{-}\ddot{C}H_2-CH=CH_2 \longleftrightarrow CH_2=CH-\ddot{C}H_2^{-}$

21. $^{-}\ddot{C}H_2-C\equiv N: \longleftrightarrow CH_2=C=\ddot{N}:^{-}$

22. $CH_3CH_2C(=O)O^{-} \longleftrightarrow CH_3CH_2C(=O)O^{-}$ (carboxylate resonance structures)

23. (cyclohexanone enolate: carbanion $\alpha$ to C=O) $\longleftrightarrow$ (cyclohexenyl oxide, $O^{-}$ on the double-bonded carbon)

24. $:N\equiv C-\ddot{C}H^{-}-CH_2-CH=CH_2 \longleftrightarrow :\ddot{N}^{-}=C=CH-CH_2-CH=CH_2$

25. no additional structures

26. no additional structures

27. (cyclohexenyl with $-\ddot{C}H_2^{-}$) $\longleftrightarrow$ (methylenecyclohexane carbanion, $=CH_2$, lone pair and negative charge on ring carbon)

28. no additional structures

29. $CH_3-\dot{C}H-CH=CH_2$

30. $(CH_3)_2\dot{C}-CH=CH-CH_3 \longleftrightarrow (CH_3)_2C=CH-\dot{C}H-CH_3$

31. (cyclohexene radical, unpaired electron on ring carbon)

32.

33.

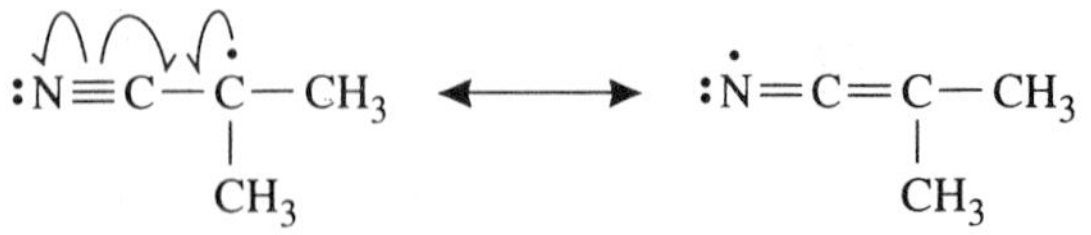

$:\!N{\equiv}C{-}\dot{C}{-}CH_3 \longleftrightarrow :\!\dot{N}{=}C{=}C{-}CH_3$

CH<sub></sub>

$CH_3$ $CH_3$

34.

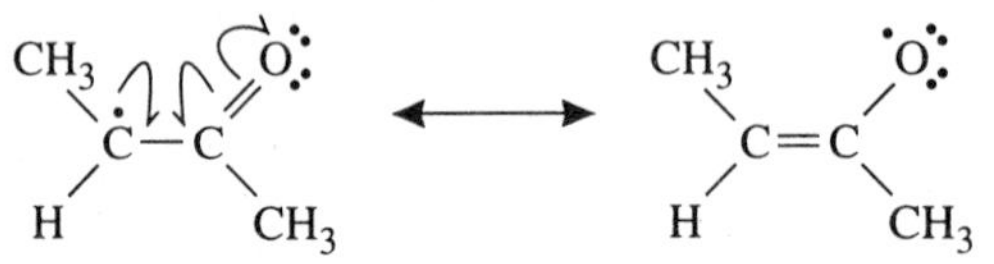

$CH_3$ $\dot{C}{-}C{=}\ddot{O}:$ H $CH_3$ $\longleftrightarrow$ $CH_3$ $C{=}C{-}\dot{O}:$ H $CH_3$

35.

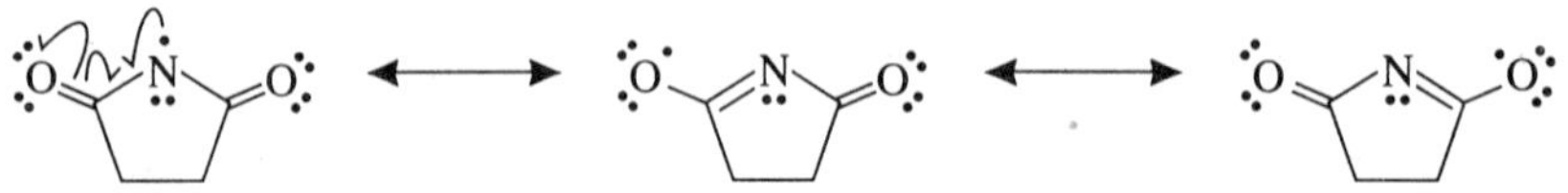

$\dot{O}{=}$ $\dot{N}$ $=\ddot{O}:$ $\longleftrightarrow$ $\dot{O}{-}$ N $=\ddot{O}:$ $\longleftrightarrow$ $\dot{O}{=}$ N $-\dot{O}:$

36.

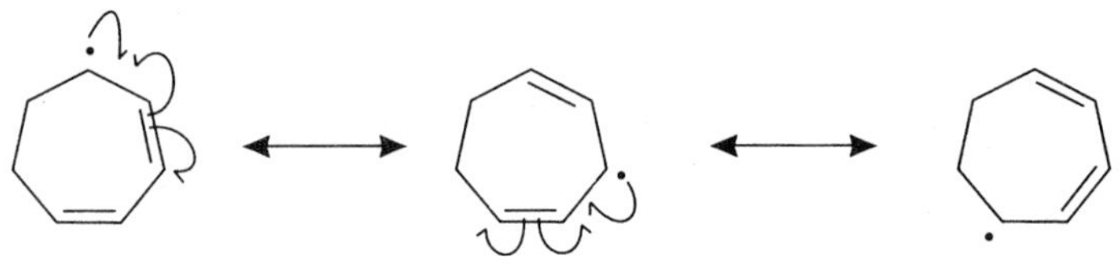

37.

38.

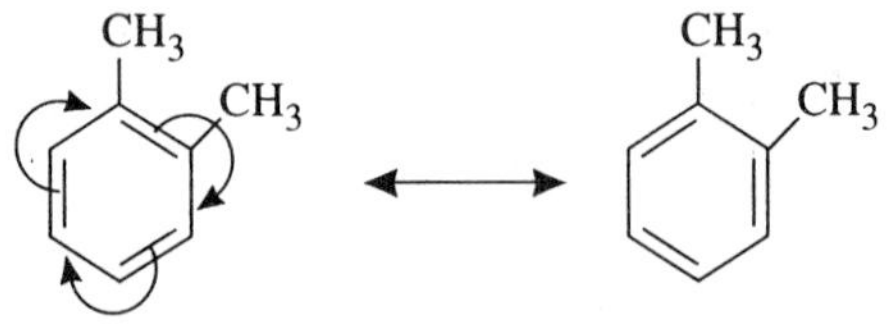

$CH_3$ $CH_3$ $\longleftrightarrow$ $CH_3$ $CH_3$

39.

CH

:ÖH

CH

:ÖH

$\longleftrightarrow$

CH

:ÖH

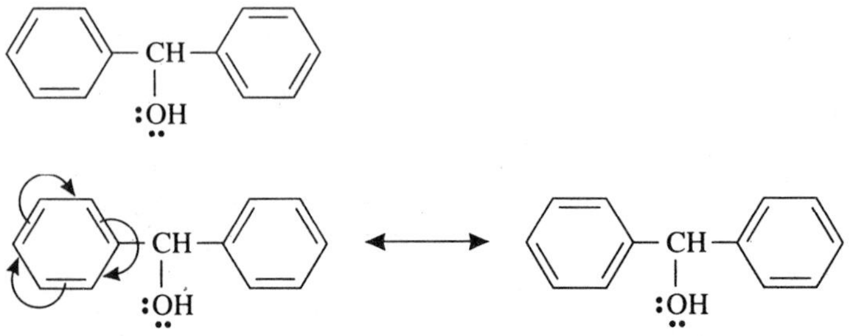

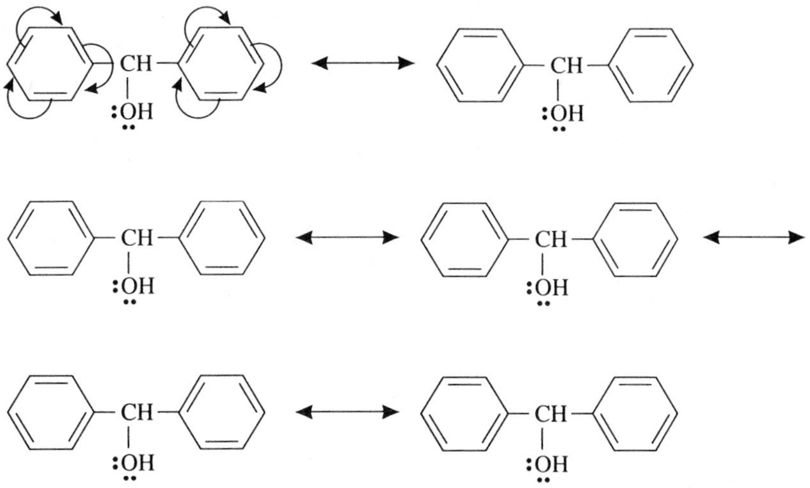

CH
:ÖH

CH
:ÖH

CH
:ÖH

CH
:ÖH

CH
:ÖH

CH
:ÖH

40.

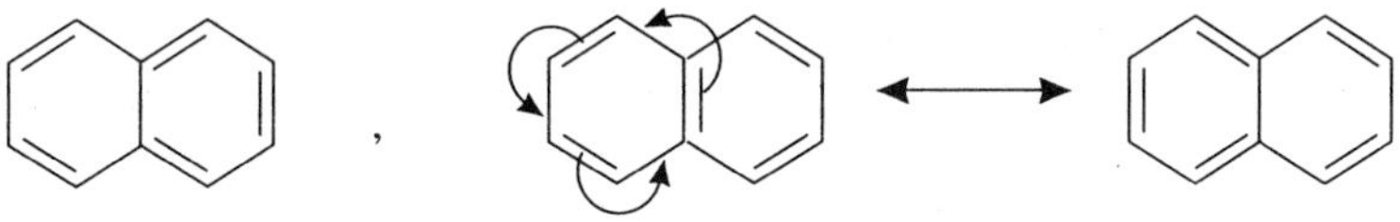

41.

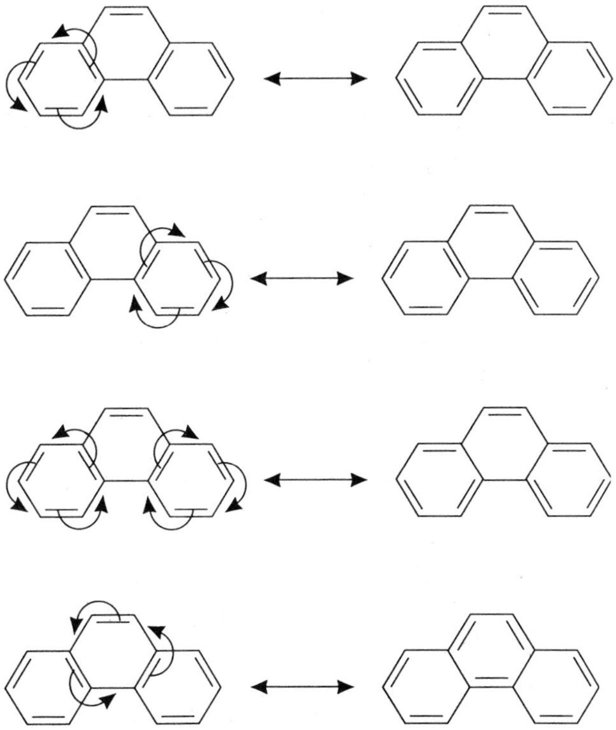

42. $CH_2-CH_3$

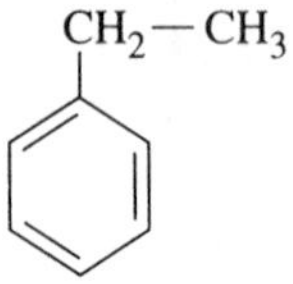

43. $CH_3$

$CH_3$

44.

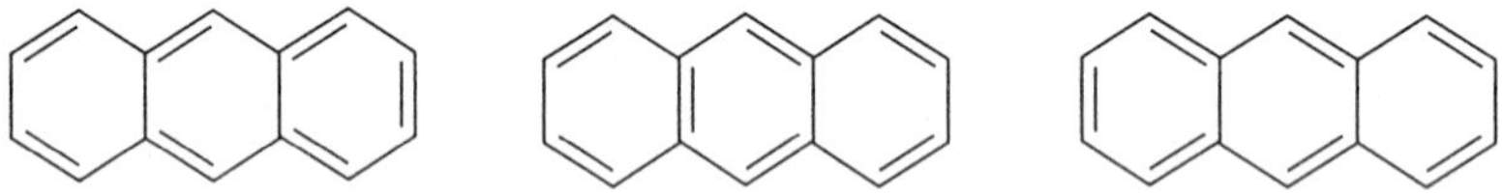

45.

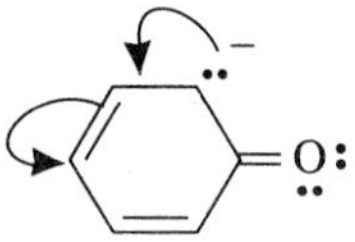

(9) 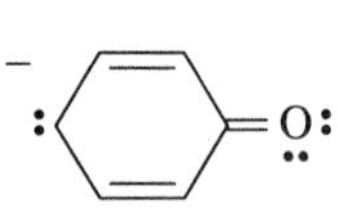 (10)

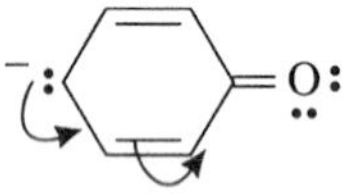

(10) 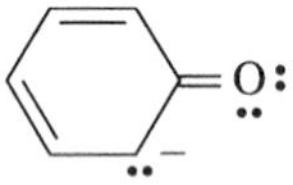 (11)

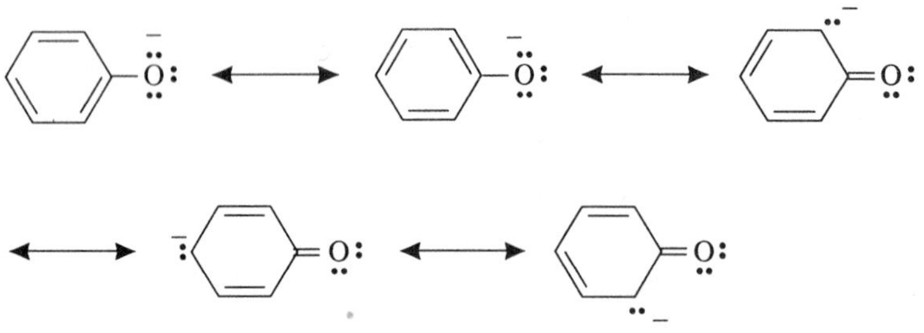

46.

(13)  (13) 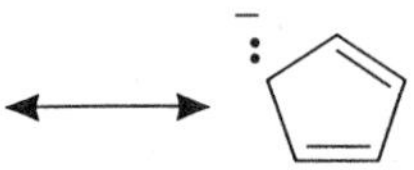 (14)

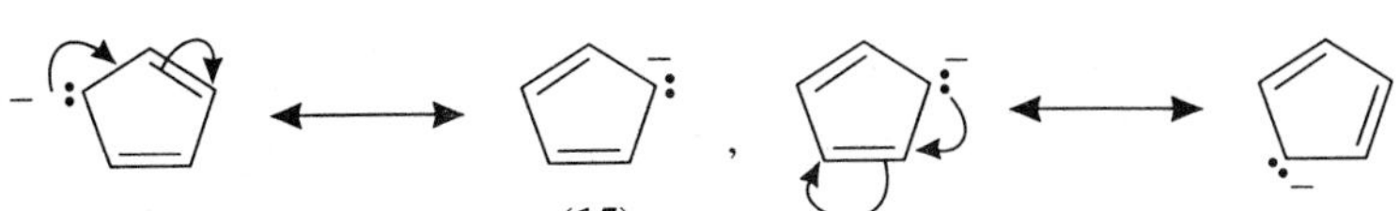

(14) ⟷ (15) , (15) ⟷ (16)

47. $\overset{+}{CH_2}$ (18) , $=CH_2$ (19) ,

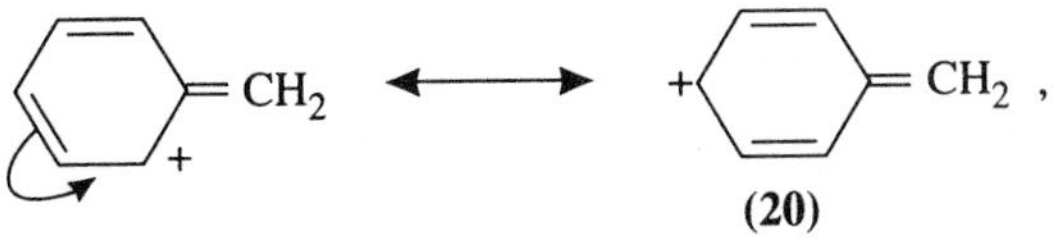

$=CH_2$ (19) ⟷ $=CH_2$ (20) ,

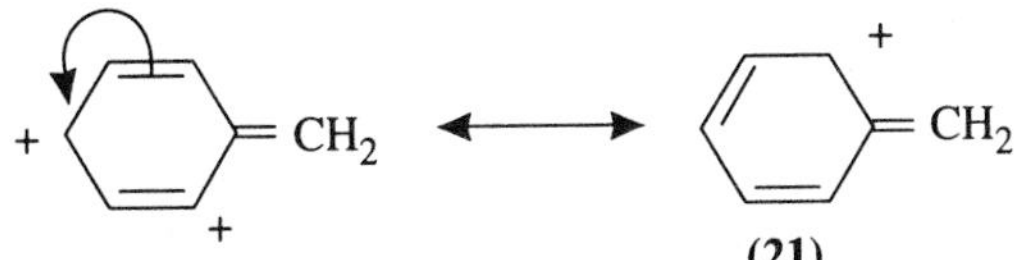

$=CH_2$ (20) ⟷ $=CH_2$ (21)

48. $\ddot{N}H_2$ H Br (22) $\ddot{N}H_2$ H Br (23) , $\ddot{N}H_2$ H Br (23) $\ddot{N}H_2$ H Br (24) ,

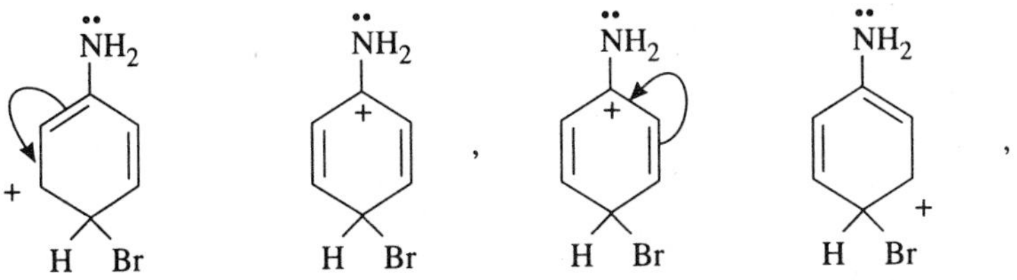

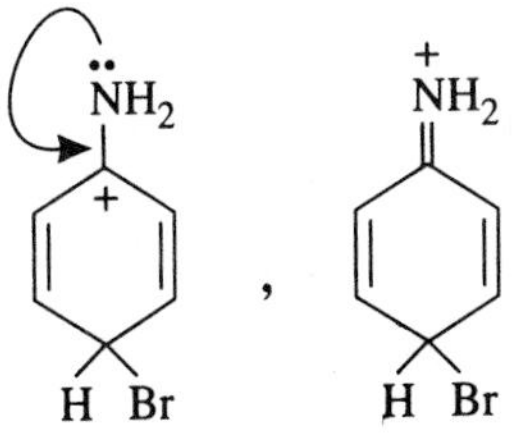

$\ddot{N}H_2$ H Br (23) , $\overset{+}{N}H_2$ H Br (25), structures (22), (23), (24), and (25)

49.

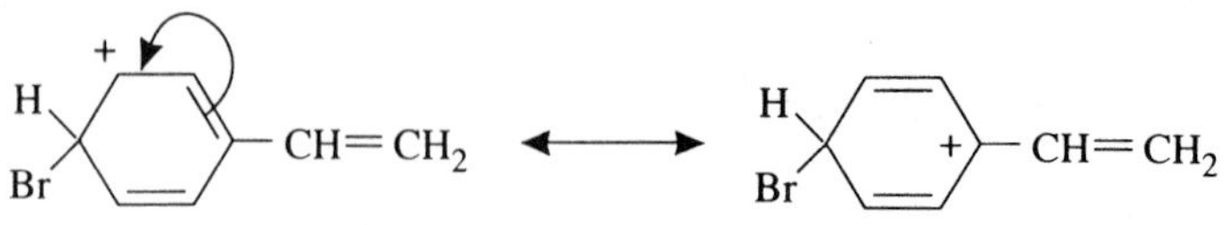

H, Br $-$CH$=$CH$_2$ ⟷ H, Br $-$CH$=$CH$_2$ ,

(26) (27)

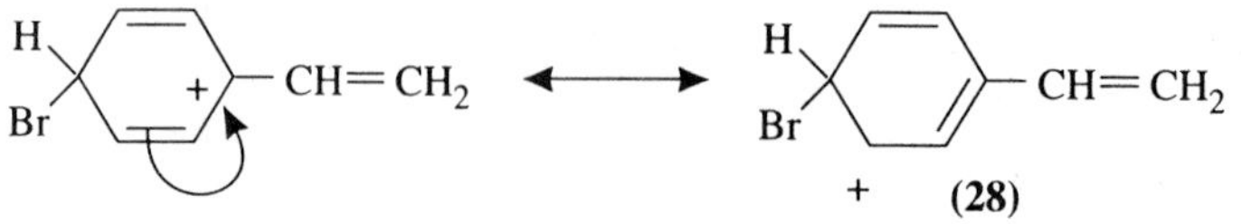

H, Br $-$CH$=$CH$_2$ ⟷ H, Br $-$CH$=$CH$_2$ ,

(27) (28)

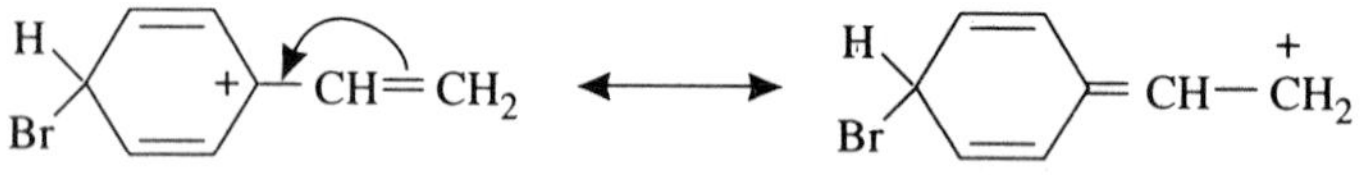

H, Br $-$CH$=$CH$_2$ ⟷ H, Br $=$CH$-$CH$_2^+$

(27) (29)

50.

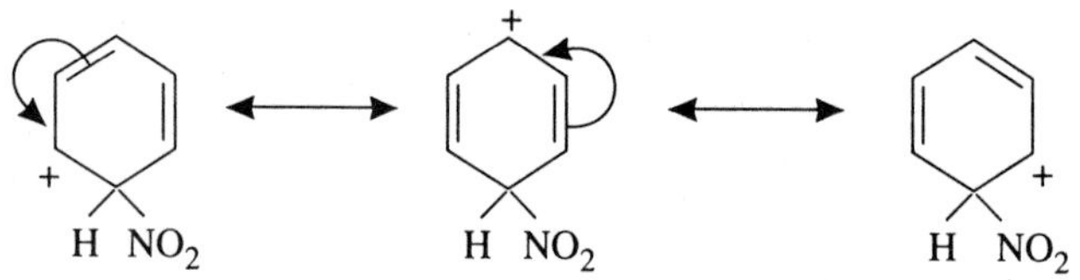

H NO$_2$ ⟷ H NO$_2$ ⟷ H NO$_2$

51.

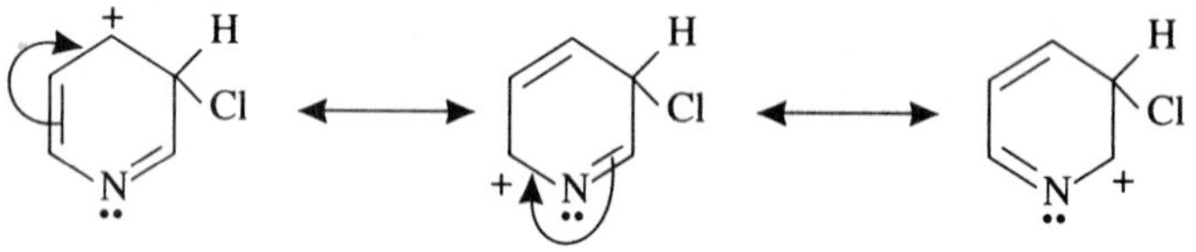

H Cl N ⟷ H Cl N ⟷ H Cl N

A fourth structure, H Cl N$^+$.

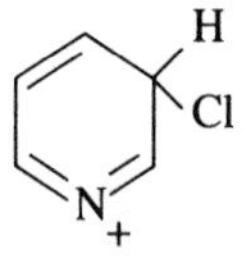

is not included because the sp-hybridized nitrogen requires that it and the two adjacent carbon atoms be linear. This adds excessive ring strain to the structure.

52.

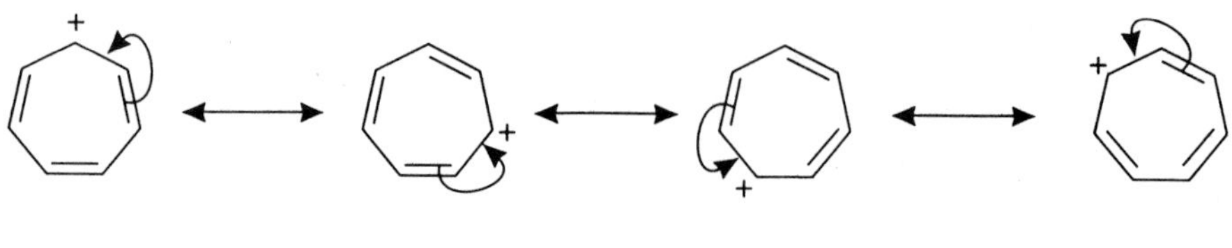

$\longleftrightarrow$ (+) $\longleftrightarrow$ (+) $\longleftrightarrow$ (+)

53. H Cl — H Cl — H Cl
$\longleftrightarrow$ $\longleftrightarrow$
N N N

54. :Ö: C N: C :O: $\longleftrightarrow$ :Ö: C N C :O: $\longleftrightarrow$ :Ö: C N: C :O

55. $CH_3$ :O: + H Cl $\longleftrightarrow$ $CH_3$ :O: + H Cl $\longleftrightarrow$ $CH_3$ :O: + H Cl $\longleftrightarrow$ $CH_3$ :O+ H Cl

56. $CH_2{=}CH{-}C(H){=}O$ , $CH_2{=}CH{-}\overset{+}{C}(H){-}O^-$ , $CH_2{=}CH{-}\overset{+}{C}(H){-}O^-$ , $\overset{+}{C}H_2{-}CH{=}C(H){-}O^-$ ,

$CH_2{=}CH{-}C(H){=}O \longleftrightarrow CH_2{=}CH{-}\overset{+}{C}(H){-}O^- \longleftrightarrow \overset{+}{C}H_2{-}CH{=}C(H){-}O^-$

57. $CH_3{-}C({=}O){-}N(CH_3)H$ , $CH_3{-}\overset{+}{C}(O^-){-}N(CH_3)H$ , $CH_3{-}\overset{+}{C}(O^-){-}N(CH_3)H$ , $CH_3{-}C(O^-){=}\overset{+}{N}(CH_3)H$ ,

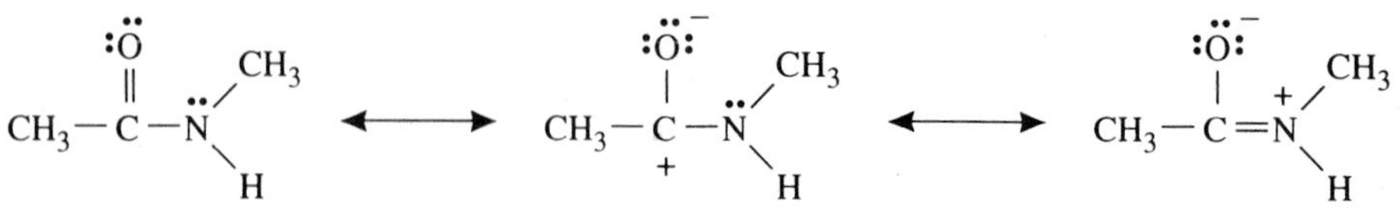

$CH_3-C-N$ ⟷ $CH_3-C-N$ ⟷ $CH_3-C=N$

58.

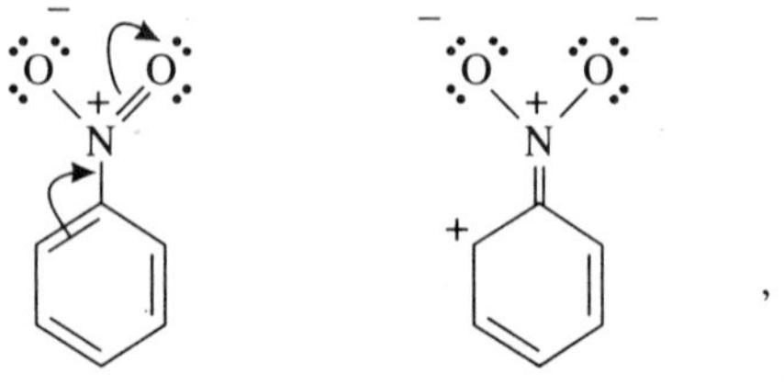

(30) (31),

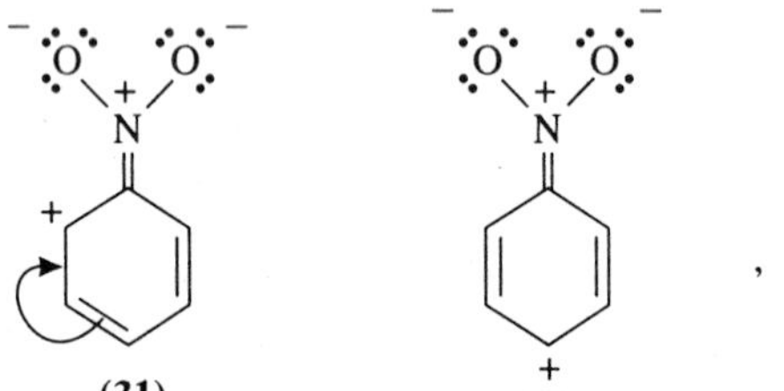

(31) (32),

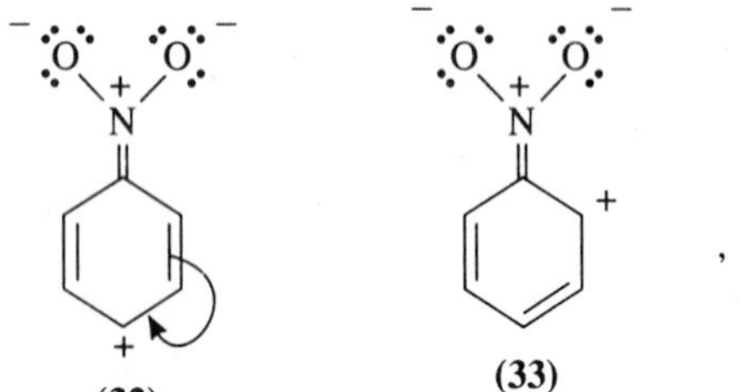

(32) (33),

Structures **(30)** through **(33)** plus

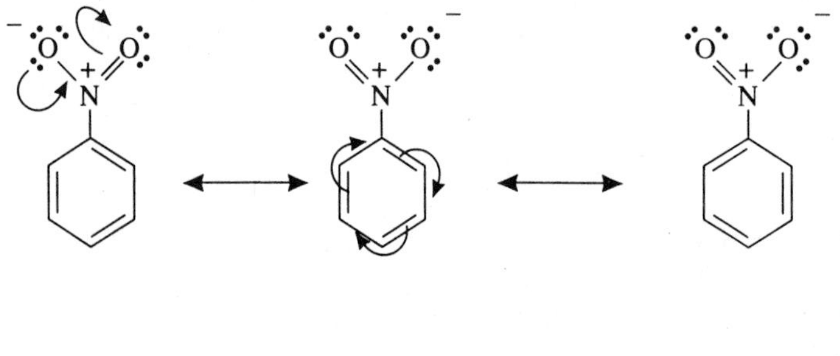

59. 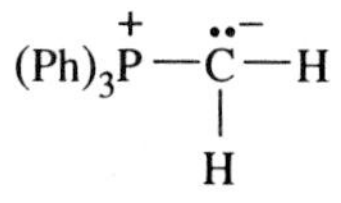 $(Ph)_3\overset{+}{P}-\overset{\cdot\cdot\,-}{C}(H)-H$

60. 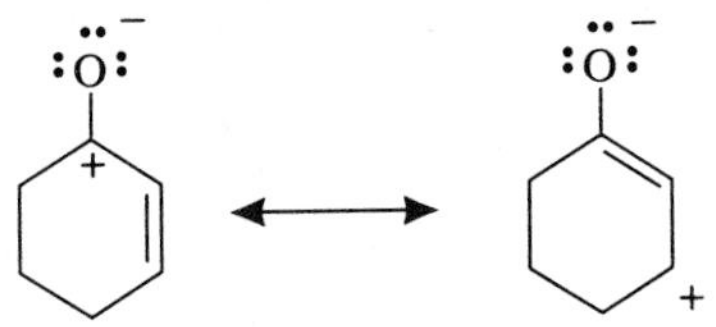 $:\overset{\cdot\cdot\,-}{O}:$ $\overset{+}{}$ $\longleftrightarrow$ $:\overset{\cdot\cdot\,-}{O}:$ $+$

61. 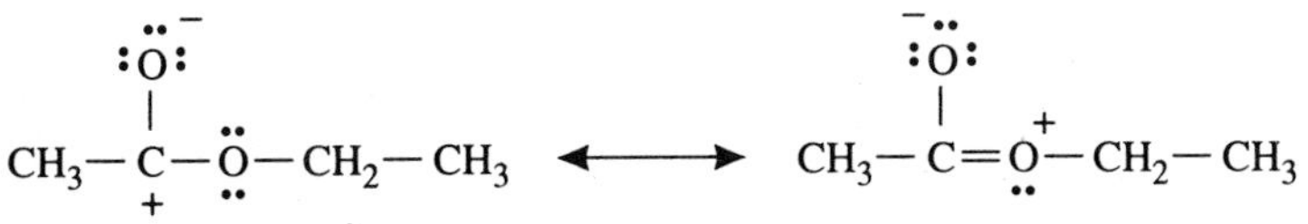 $CH_3-\overset{+}{C}(:\overset{\cdot\cdot\,-}{O}:)-\ddot{O}-CH_2-CH_3 \longleftrightarrow CH_3-C(:\overset{-\,\cdot\cdot}{O}:)=\overset{+}{O}-CH_2-CH_3$

62. 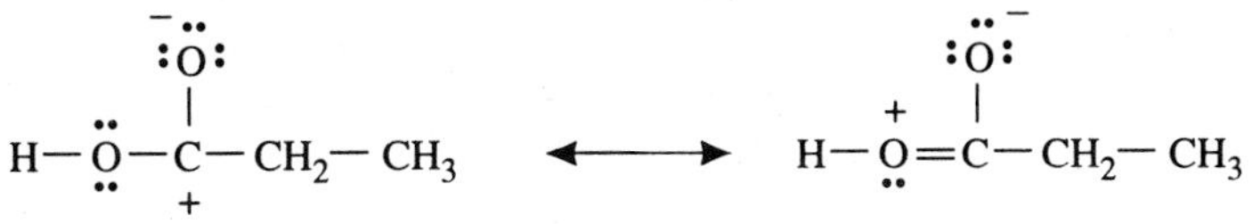 $H-\ddot{O}-\overset{+}{C}(:\overset{-\,\cdot\cdot}{O}:)-CH_2-CH_3 \longleftrightarrow H-\overset{+}{O}=C(:\overset{\cdot\cdot\,-}{O}:)-CH_2-CH_3$

63. 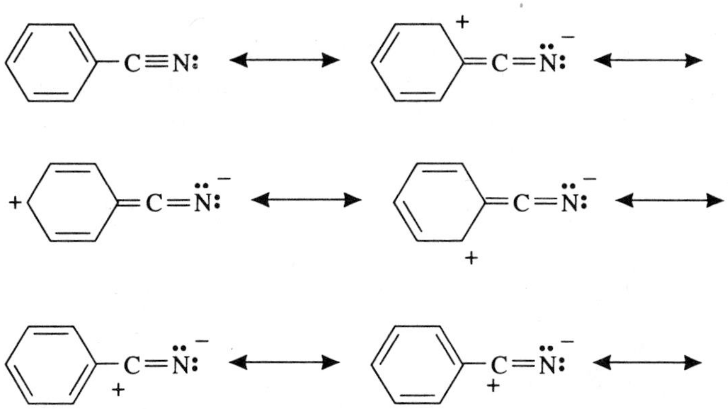 $-C\equiv N: \longleftrightarrow {}^{+}\ =C=\overset{\cdot\cdot\,-}{N}: \longleftrightarrow$

$+\ =C=\overset{\cdot\cdot\,-}{N}: \longleftrightarrow\ =C=\overset{\cdot\cdot\,-}{N}:\ + \longleftrightarrow$

$-\overset{+}{C}=\overset{\cdot\cdot\,-}{N}: \longleftrightarrow -\overset{+}{C}=\overset{\cdot\cdot\,-}{N}: \longleftrightarrow$

64. 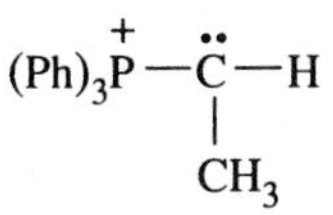 $(Ph)_3\overset{+}{P}-\overset{\cdot\cdot}{C}(CH_3)-H$

# 3 Mechanisms

A mechanism is a step-by-step description of how a chemical reaction occurs. Each step will involve some kind of bond breaking and/or bond making. Pushing electrons is an excellent, graphic way to describe each step.

Development of experimental methods to elucidate mechanisms is one of the triumphs of science and was led by organic chemists. Thanks to the creativity and skill exercised by physical organic chemists over the last six decades, students can better understand how products arise from starting materials. Further, students can organize an otherwise intimidating amount of factual content into categories of mechanisms making the study of organic chemistry much more manageable. Chemists use mechanisms to predict new reactions as well as to understand old reactions.

In Chapter 1 we briefly trespassed into mechanisms when we pushed electrons to make ions and free radicals. In Chapter 2 you learned to push electrons in order to generate resonance structures. Now it is time to become an accomplished electron pusher. You can never be too skilled in pushing electrons.

## Sigma Bond Breaking

Heterolytic cleavage of **sigma bonds** occurs under a variety of conditions. It is shown below. The arrow indicates that the sigma electrons that form the A–B bond are leaving A and becoming the exclusive property of B. Since the fragment A is formally losing one electron, it must become positively charged and B must become negatively charged since it gains an electron.

$$A{-}\ddot{\underset{\cdot\cdot}{B}}{:} \longrightarrow A^+ + {:}\ddot{\underset{\cdot\cdot}{B}}{:}^-$$

This occurs in the first step of the **familiar $S_N1$ reaction.** In *t*-butyl chloride, the leaving group is *chloride ion.* The *carbon–chlorine* bond cleaves, the electrons going with *chlorine,* which obtains a *negative* charge. The carbon atom obtains a *positive* charge. Thus, the cleavage of *t*-butyl chloride yields chloride ion plus the *t*-butyl cation.

$$CH_3-\underset{CH_3}{\overset{CH_3}{C}}-\ddot{\underset{..}{Cl}}: \longrightarrow CH_3-\underset{CH_3}{\overset{CH_3}{C^+}} + :\ddot{\underset{..}{Cl}}:^-$$

1. In isopropyl chloride the leaving group is __________ ion. The __________ – __________ bond cleaves, the electrons going with __________, that obtains a __________ charge. The carbon atom obtains a __________ charge. Thus, the cleavage of isopropyl chloride yields chloride ion plus the __________ cation.

$$CH_3-\underset{H}{\overset{CH_3}{C}}-\ddot{\underset{..}{Cl}}: \longrightarrow \underline{\qquad\qquad\qquad}$$

2. In cyclohexyl bromide the leaving group is __________ ion. The __________ – __________ bond cleaves, the electrons going with __________, that obtains a __________ charge. The carbon atom obtains a __________ charge. Thus, the cleavage of cyclohexyl bromide yields __________ ion plus the __________ cation.

$$C_6H_{11}-\ddot{\underset{..}{Br}}: \qquad \longrightarrow$$

______________________ → ______________________

supply arrow

* * * * *

Very often, heterolytic cleavage occurs from a charged intermediate which was formed in a previous step. The process is the same, but the charge on the products is different.

$$A-\overset{+}{\underset{..}{\ddot{B}}}-H \longrightarrow A^+ + :\ddot{\underset{..}{B}}-H$$

3. This kind of cleavage occurs in the second step of the dehydration of alcohols. In the first step the alcohol is protonated to form (1). In (1), the leaving group is a *water* molecule. The *carbon–oxygen* bond cleaves, the electrons going with the oxygen atom. The *oxygen* atom, which had

a *positive* charge, becomes *neutral.* The carbon atom obtains a *positive* charge. Thus, the cleavage of (1) yields *water* plus the *t-butyl* cation.

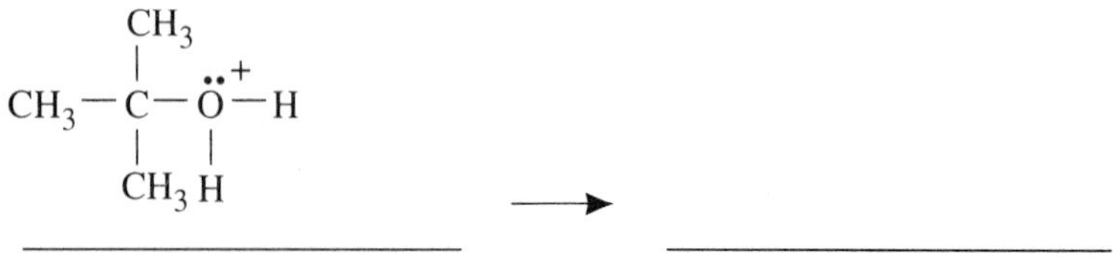

$$CH_3-\underset{CH_3}{\overset{CH_3}{\underset{|}{\overset{|}{C}}}}-\underset{H}{\underset{|}{\ddot{O}^+}}-H \longrightarrow$$

______________________ ______________________

supply arrow

**(1)**

4. The acid-catalyzed hydrolysis of methoxymethyl acetate begins with protonation giving (2). A C–O bond cleaves, the electrons going with the __________ atom. The oxygen atom that had a positive charge becomes __________. The leaving group is *acetic acid.* The carbon atom from which acetic acid departed obtains a __________ charge. Thus, the cleavage of (2) yields acetic acid plus the *methoxymethyl* cation.

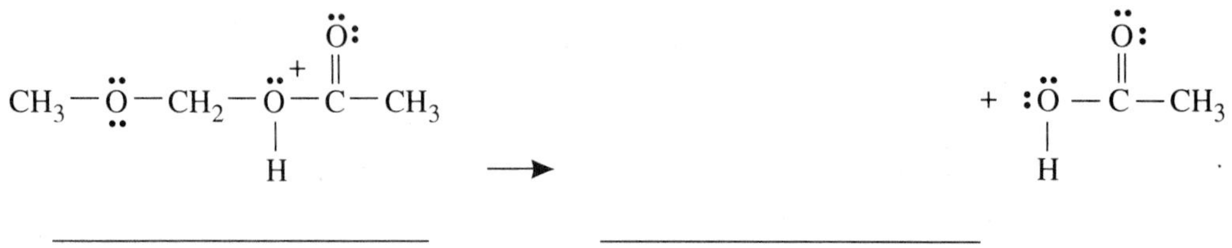

$$CH_3-\ddot{\underset{..}{O}}-CH_2-\underset{H}{\underset{|}{\ddot{O}^+}}-\overset{\ddot{O}:}{\overset{\|}{C}}-CH_3 \longrightarrow \qquad + \; :\underset{H}{\underset{|}{\ddot{O}}}-\overset{\ddot{O}:}{\overset{\|}{C}}-CH_3$$

______________________ ______________________

supply arrow

**(2)**

5. Sometimes a ring will be opened as a result of sigma bond breaking. The **protonated** epoxide, (3), will suffer heterolytic cleavage of one of its carbon–oxygen bonds, with the electrons going to the __________ atom. The oxygen atom, which had a __________ charge, becomes __________. The leaving group is

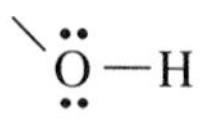

$$\diagdown\ddot{\underset{..}{O}}-H$$

The carbon atom from which the oxygen departed obtains a __________ charge.

$$\text{epoxide ring: } (CH_3)_2C-C(CH_3)_2 \text{ bridged by } \ddot{O}^+-H \longrightarrow CH_3-\overset{CH_3}{\overset{|}{C}}(-\ddot{\underset{..}{O}}-H)-\underset{CH_3}{\underset{|}{C^+}}-CH_3$$

______________________ ______________________

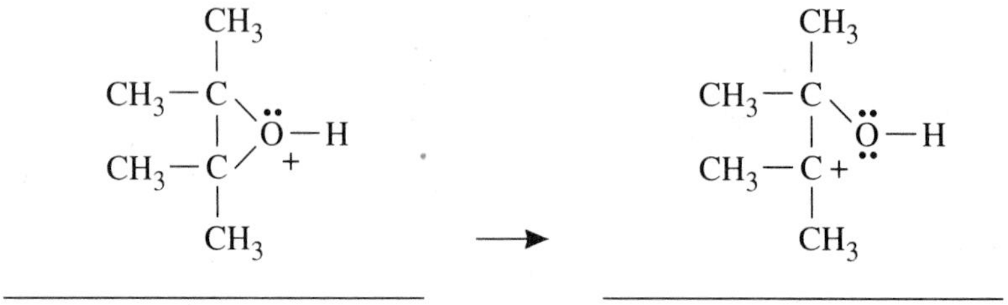

supply arrow

**(3)**

## Exercises

Here are some exercises in sigma bond breaking. Supply the arrows and the products. Remember that charge must be conserved. That is, if a neutral molecule is dissociated, the algebraic sum of the charges on the products must equal zero. If a positively charged ion is dissociated, the algebraic sum of the charges on the products must equal +1.

6. $C_6H_5-CH_2-\ddot{Br}:$ ⟶ ______

7. $:\ddot{Cl}-CH(C_6H_5)-C_6H_5$ ⟶ ______

8. Cyclohexyl(H)$-\ddot{O}^{+}(H)-SO_2-C_6H_4-CH_3$ ⟶ ______

9. $CH_3-\ddot{O}-C_6H_4-CH_2-\ddot{Cl}:$ ⟶ ______

10. $(Ph)_2C-C(Ph)_2$ bridged by $\ddot{O}^{+}-H$ ⟶ ______

11. $C_6H_5-C(=\ddot{O}:)-\ddot{O}^{+}(H)-CH_2-\ddot{O}-CH_2-CH_3$ ⟶ ______

12.

$$\begin{array}{c} CH_3 \\ | \\ H-\ddot{O}^{+}\!\!<\begin{array}{c} CH \\ | \\ CH \end{array} \\ | \\ CH_3 \end{array}$$

$\longrightarrow$

13.

$$CH_3-\underset{CH_3}{\overset{CH_3}{C}}-\underset{H}{\overset{+}{\ddot{O}}}-\overset{\ddot{O}:}{\overset{\|}{C}}-C_6H_5$$

$\longrightarrow$

14.

$$H-\underset{+}{\overset{SO_2-C_6H_4-\ddot{B}r:}{\ddot{O}}}-C_5H_9$$

$\longrightarrow$

## Sigma Bond Making

Formation of a sigma bond occurs when an anion and a cation encounter one another. That is

$$A^{+} \curvearrowleft :\ddot{B}:^{-} \longrightarrow A-\ddot{B}:$$

The arrow indicates that a pair of electrons which was the exclusive property of B is now shared by A and B.

The product-forming step in many $S_N1$ reactions involves the formation of a sigma bond between the intermediate carbocation and some negatively charged nucleophile. For example, α-phenylethyl cation will yield an ester by reaction with acetate ion. A pair of unshared electrons on the negatively charged *oxygen* atom forms a bond between *oxygen* and *carbon* and becomes a shared pair of sigma electrons.

The oxygen atom that had a *negative* charge becomes *neutral,* and the carbon atom that had a *positive* charge becomes *neutral.*

$$\text{Ph}-\overset{+}{\text{C}}\text{H}(\text{CH}_3) + {}^{-}\text{:}\ddot{\text{O}}-\text{C}(=\ddot{\text{O}}\text{:})-\text{CH}_3 \longrightarrow \text{Ph}-\text{CH}(\text{CH}_3)-\ddot{\text{O}}-\text{C}(=\ddot{\text{O}}\text{:})-\text{CH}_3$$

15. *Sec*-butyl cation will react with bromide ion to form *sec*-butyl bromide. A pair of ______________ electrons on the negatively charged ______________ ion forms a bond between ______________ and ______________ and becomes a pair of shared sigma electrons. The bromine atom that had a ______________ charge becomes ______________.

$$\text{CH}_3-\text{CH}_2-\overset{+}{\text{C}}\text{H}(\text{CH}_3) \qquad \text{:}\ddot{\text{B}}\text{r}\text{:}^{-} \longrightarrow$$

______________________________ ______________________________

supply arrow

* * * * *

More commonly, sigma bonds are formed by the reaction of carbocations with neutral molecules possessing unshared pairs. The result is a positively charged intermediate.

$$-\overset{|}{\underset{|}{\text{C}}}_{+} \quad \ddot{\text{N}}\text{u} \longrightarrow -\overset{|}{\underset{|}{\text{C}}}-\text{Nu}^{+}$$

16. During the solvolysis of alkyl halides in ethanol, a carbocation forms a sigma bond with one of the unshared pairs on the alcohol oxygen atom. A pair of unshared electrons on the oxygen atom forms a bond between *oxygen* and *carbon* and becomes a shared pair of *sigma* electrons. The oxygen atom, which was *neutral,* now possesses a formal *positive* charge. The carbon atom, which had a *positive* charge, is now *neutral.*

$$\text{Ph}-\text{C}^{+}\text{H}(\text{CH}_3) \qquad \text{:}\ddot{\text{O}}(\text{H})-\text{CH}_2-\text{CH}_3 \longrightarrow \text{Ph}-\text{CH}(\text{CH}_3)-\overset{+}{\ddot{\text{O}}}(\text{H})-\text{CH}_2-\text{CH}_3$$

______________________________

supply arrow

17. Carbocations react with amines in an analogous process. The __________ electrons on the nitrogen atom form a bond between __________ and __________ and becomes a pair of __________ electrons. The nitrogen atom, which was neutral, now possesses a formal __________ charge. The carbon atom which had a __________ charge is now __________.

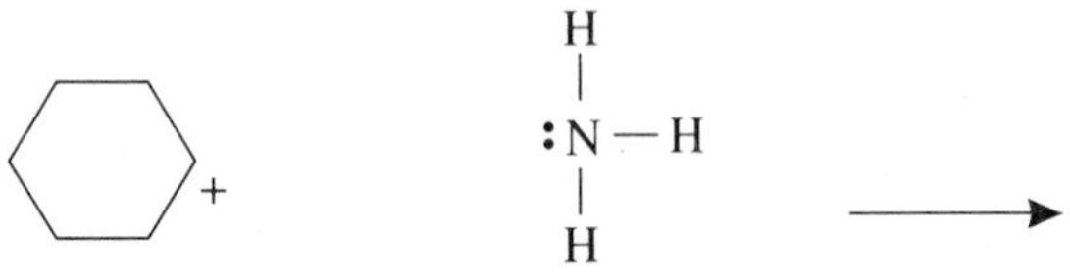

$$\text{(cyclohexyl cation)}^{+} \quad :NH_3 \longrightarrow$$

______________________ ______________________

supply arrow

* * * * *

Note some important features of the way electron pushing is used to write mechanisms.

1. **Electrons are always pushed *toward* a center of positive charge.**
2. **Although the position of positive charge changes, the positive charge is *not* pushed. Arrows should never lead away from a positive charge.**
3. **Total charge is conserved. That is, the sum of the charges on one side of the equation always equals the sum of the charges on the other.**
4. **Electrons are conserved. That is, the number of valence electrons on each side of an equation is always equal.**

## Exercises

Here are some examples of sigma bond making. Supply the arrows and the products.

18. $C_6H_5-\overset{+}{C}H_2 \quad :\ddot{\underset{..}{F}}:^{-} \longrightarrow$

______________________ ______________________

19. (cyclohexyl cation, $^{+}$C–H) $\quad ^{-}:\ddot{\underset{..}{O}}-\overset{\overset{:\ddot{O}}{\|}}{C}-CH_3 \longrightarrow$

______________________ ______________________

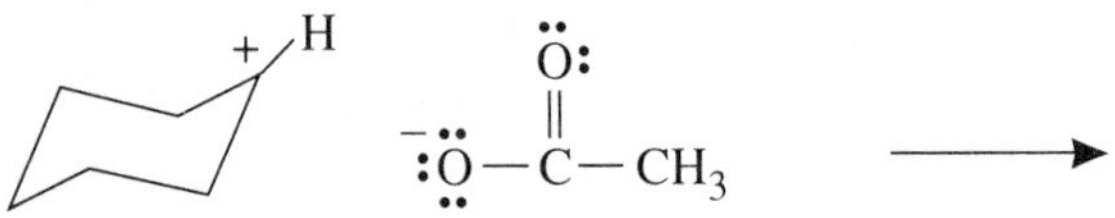

20. $CH_3-CH_2-\overset{\overset{CH_3}{|}}{\underset{+}{C}H} \quad :\ddot{O}(H)-CH_3 \longrightarrow$

______________________ ______________________

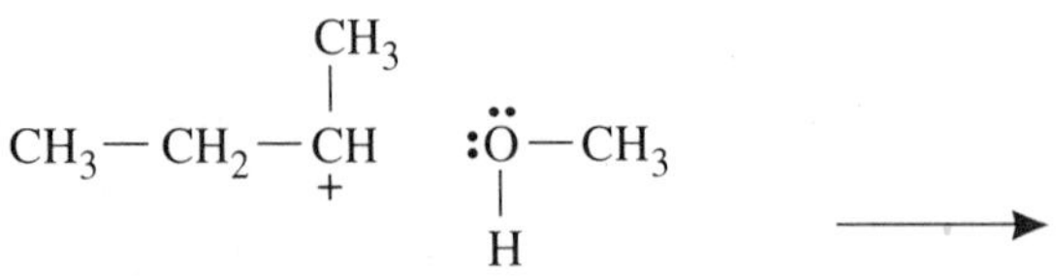

21.

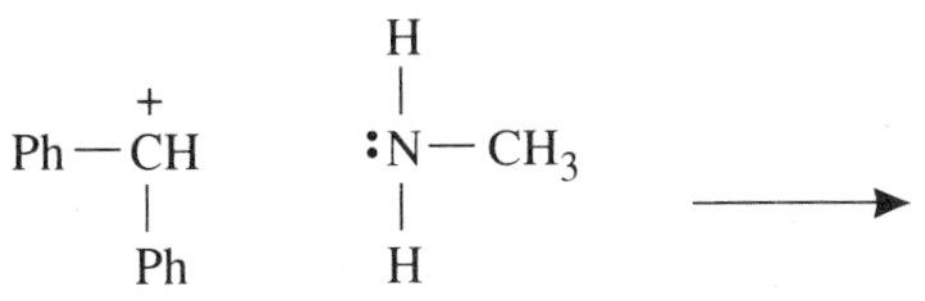

$$Ph-\overset{+}{C}H(-Ph) \quad :N(H)(H)-CH_3 \longrightarrow$$

22.

$$\overset{}{CH_2}-\overset{+}{CH_2} \quad H-\ddot{O}: \longrightarrow$$

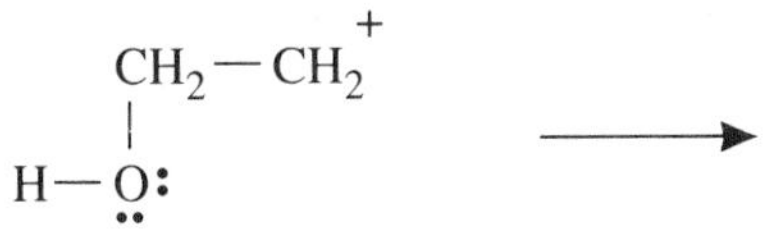

# Simultaneous Bond Making and Breaking

While isolated events of bond breaking and bond making are important in writing mechanisms of organic reactions, it is far more common to encounter steps in which bond making and bond breaking occur simultaneously. In some instances, bond making has proceeded well along toward completion and bond breaking has only begun as the transition state is reached. The opposite also occurs. These are fine points of mechanism and need not concern a student just learning to push electrons.

## Sigma Bond Making and Breaking

The simultaneous making and breaking of sigma bonds is illustrated nicely in the familiar **$S_N2$ reaction.** A negatively charged nucleophile, $\ddot{N}u^-$, approaches a carbon atom having a leaving group, –L, in a direction anti to and rearward to the leaving group. The pushable electrons are possessed by the nucleophile and the receptor is the carbon atom that itself has a pair of pushable electrons, namely, the sigma electrons forming the C–L bond. The products are a compound having a C–Nu bond and the anion of the leaving group. Thus, Nu has been substituted for L.

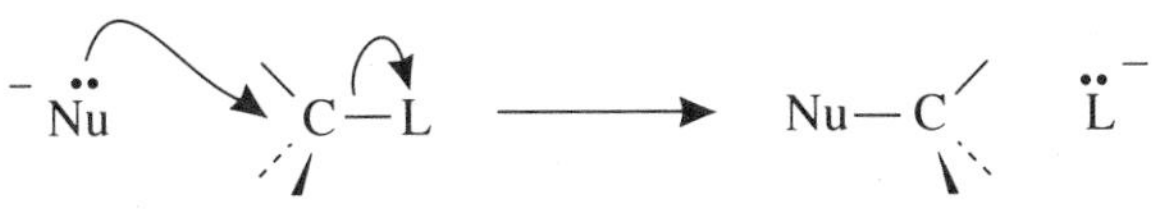

$$^-\ddot{N}u \quad C-L \longrightarrow Nu-C \quad \ddot{L}^-$$

This illustrative equation has been written in the **stereochemically** accurate form showing that an inversion of configuration occurs during $S_N2$ reactions. It was done this way to make the illustration as clear as possible. However, the student should note that the $S_N2$ reaction is usually written without reference to stereochemistry unless a stereochemical point is being made. Thus, the reaction is usually written

$$^-\ddot{N}u \quad -\overset{|}{\underset{|}{C}}-L \longrightarrow Nu-\overset{|}{\underset{|}{C}}- \quad \ddot{L}^-$$

The $S_N2$ reaction of methyl iodide with bromide ion illustrates the process. The nucleophile is *bromide ion.* The leaving group is *iodide ion.* A pair of unshared electrons on the bromide ion is pushed toward the carbon atom and, simultaneously, the ***sigma* electrons** of the carbon–iodine bond are pushed toward *iodine.* The result is the making of a sigma bond between bromine and carbon, and the breaking of a sigma bond between carbon and iodine. The bromine, which had a *negative* charge, becomes *neutral* and the iodine, which was *neutral,* now possesses a formal *negative* charge.

$$:\ddot{\underset{..}{Br}}:^{-} \quad CH_3-\ddot{\underset{..}{I}}: \longrightarrow :\ddot{\underset{..}{Br}}-CH_3 \quad :\ddot{\underset{..}{I}}:^{-}$$

23. Another example is the reaction of hydroxide ion with ethyl chloride. The nucleophile is ______________ ion. The leaving group is ______________ ion. A pair of ______________ electrons from hydroxide ion is pushed toward the carbon atom and, simultaneously, the ______________ electrons of the carbon–chlorine bond are pushed toward ______________. The result is the making of a sigma bond between oxygen and ______________ and the breaking of the sigma bond between carbon and ______________. The oxygen, which had a negative charge, is now ______________ and the chlorine, which was neutral, now possesses a formal ______________ charge.

$$H-\ddot{\underset{..}{O}}:^{-} \quad CH_3CH_2-\ddot{\underset{..}{Cl}}: \longrightarrow$$

______________________ ______________________

supply arrows

24. Consider the reaction of hydroxide ion with benzyl bromide.

$$H-\ddot{\underset{..}{O}}:^{-} \quad C_6H_5CH_2-\ddot{\underset{..}{Br}}: \longrightarrow$$

______________________ ______________________

supply arrows

* * * * *

Carbon need not always be the center of this type of reaction. The removal of an acidic proton from an organic compound by a base is a common occurrence.

$$A-H \quad :B^{-} \longrightarrow \ddot{A}^{-} \quad H-B$$

25. The reaction of phenol with hydroxide ion results in the phenoxide ion plus a molecule of water.

A pair of unshared electrons on the oxygen atom of the hydroxide ion is pushed toward the hydrogen atom and, simultaneously, the *sigma* electrons of the bond between the hydrogen and oxygen atoms of phenol are pushed toward *oxygen.* The result is the making of a sigma bond between oxygen and ________________ and the breaking of a sigma bond between hydrogen and ________________ The oxygen atom of the hydroxide ion, which had a *negative* charge, is now *neutral,* and the oxygen atom of phenol, which was *neutral,* now possesses a formal *negative* charge.

$$C_6H_5-\ddot{\underset{\cdot\cdot}{O}}-H \quad :\ddot{\underset{\cdot\cdot}{O}}^{-}-H \longrightarrow \underline{\qquad\qquad\qquad}$$

26. The amide ion is a strong enough base to remove a proton from acetylene. The reaction results in a molecule of ammonia plus the acetylide anion. A pair of ________________ electrons on the ________________ atom of the amide ion is pushed toward the hydrogen atom, and, simultaneously, the ________________ electrons of the carbon–hydrogen bond of acetylene are pushed toward ________________ Thus, a sigma bond is made between nitrogen and ________________, and a sigma bond is broken between hydrogen and ________________ The nitrogen atom that possessed a negative charge in the amide ion is now ________________, and the carbon atom that was neutral in acetylene now possesses a ________________ charge.

$$H-C\equiv C-H \quad :\ddot{N}^{-}(H)-H \longrightarrow$$

________________ ________________

supply arrows

27. The reaction of ethoxide ion with one of the alpha hydrogens of acetaldehyde is similar. The products are a molecule of ethanol and the anion of acetaldehyde.

$$CH_3-CH_2-\ddot{\underset{\cdot\cdot}{O}}:^{-} \quad H-CH_2-C(=\ddot{O}:)-H \longrightarrow$$

________________ ________________

supply arrows

* * * * *

The formation of a carbon–carbon sigma bond occurs in the reaction of a Grignard reagent with ethylene oxide. In the illustration the **Grignard reagent,** ethyl magnesium bromide, is written as an ion pair consisting of the $MgBr^+$ ion and the ethyl anion. The nucleophile is the *ethyl anion.* A pair of unshared electrons on the carbanion is pushed toward one of the *carbon* atoms of ethylene oxide. Simultaneously, the *sigma* electrons of the carbon–oxygen bond are pushed toward *oxygen.* The result is a ring opening. The carbanion, which had a *negative* charge, has become *neutral* and the oxygen atom of ethylene oxide, which was *neutral,* now possesses a formal *negative* charge.

$$CH_3-\overset{-\,\cdot\cdot}{C}H_2 \;\; \overset{+}{MgBr} \quad + \quad \text{(ethylene oxide: } CH_2-CH_2 \text{ ring with } \ddot{O}\text{)} \longrightarrow \overset{+}{MgBr} \quad CH_3-CH_2-CH_2-CH_2-\ddot{O}^{-}$$

28. In the reaction of propyl magnesium iodide with ethylene oxide, the nucleophile is the ________________ anion. A pair of ________________ electrons on the carbanion is pushed toward one of the ________________ atoms of ethylene oxide. Simultaneously, the ________________ electrons of the carbon– ________________ bond are pushed toward ________________. Thus, a sigma bond has been made between carbon and ________________, and a sigma bond has been broken between carbon and ________________. The carbanion, which had a negative charge, has become ________________, and the oxygen atom of ethylene oxide, which was neutral, now possesses a ________________ charge.

$$CH_3-CH_2-\overset{-\,\cdot\cdot}{C}H_2 \;\; \overset{+}{MgI} \quad \text{(ethylene oxide: } CH_2-CH_2 \text{ ring with } \ddot{O}\text{)} \longrightarrow$$

________________________ ________________________

supply arrows

29. Analogously, an alcoholate anion will open an epoxide ring. In the reaction of ethoxide ion with 2-butene oxide, the nucleophile is ________________ ion. A pair of ________________ electrons on the ________________ atom of the ethoxide ion is pushed toward one of the ________________ atoms in 2-butene oxide. Simultaneously, the ________________ electrons of the carbon–________________ bond are pushed toward ________________. Thus, a sigma bond has been made between oxygen and ________________, and a sigma bond has been broken between carbon and ________________. The ethoxide ion, which had a negative charge, has become ________________, and the oxygen atom of 2-butene oxide, which was neutral, now possesses a ________________ charge.

$$CH_3-CH_2-\ddot{\underset{\cdot\cdot}{O}}{:}^{-} \quad \text{(2-butene oxide: } CH_3-CH-CH-CH_3 \text{, CH–CH ring with } \ddot{O}\text{)} \longrightarrow$$

________________________ ________________________

supply arrows

* * * * *

The simultaneous making and breaking of sigma bonds occurs also between a neutral nucleophile or base with pushable electrons and a positively charged species that can act as a receptor.

$$\ddot{Nu} \quad A{-}X^{+} \longrightarrow Nu{-}A^{+} \quad + \ddot{X}$$

In organic chemistry the most common examples of this are reactions in which a molecule possessing unshared electrons is protonated by a positively charged acid.

Alcohols are protonated by hydronium ion ($H_3O^+$). In the reaction of *t*-butyl alcohol with hydronium ion the base is *t*-butyl alcohol, which has unshared pairs on the *oxygen* atom. The acid is *hydronium ion.* The unshared electrons on the oxygen atom of *t*-butyl alcohol are pushed toward the *hydrogen* atom. Simultaneously, the pair of *sigma* electrons between the *hydrogen* and *oxygen* atoms of the hydronium ion is pushed toward the *oxygen* atom. Thus, an *oxygen–hydrogen* sigma bond is made and a *hydrogen–oxygen* sigma bond is broken. The oxygen atom of the alcohol, which was *neutral,* now possesses a formal *positive* charge and the oxygen atom of the hydronium ion, which had a *positive* charge, is now *neutral.*

$$(CH_3)_3C{-}\ddot{O}{:}{-}H \quad + \quad H{-}\ddot{O}^{+}H_2 \longrightarrow (CH_3)_3C{-}\ddot{O}^{+}H_2 \quad + \quad {:}\ddot{O}H_2$$

30. Cyclohexanol can be protonated by hydronium ion. The base is ________________, which has unshared pairs of electrons on the ________________ atom. The acid is ________________ ion. A pair of ________________ electrons on the oxygen atom of cyclohexanol is pushed toward the ________________ atom. Simultaneously, the pair of ________________ electrons between the hydrogen and ________________ atoms of the hydronium ion is pushed toward the ________________ atom. Thus, an oxygen–________________ sigma bond is made, and a hydrogen– ________________ sigma bond is broken. The oxygen atom of the alcohol, which was neutral, now possesses a ________________ charge, and the oxygen atom of the former hydronium ion, which had a positive charge, is now ________________.

$$C_6H_{11}{-}\ddot{O}{-}H \quad + \quad H_3\ddot{O}^{+} \longrightarrow$$

________________________________ ________________________________

supply arrows

31. The reaction just described is reversible. Deprotonation of the conjugate acid of an organic base

by water provides another example of simultaneous making and breaking of sigma bonds. Thus, in the deprotonation of anilinium ion by water, the base is water, which has unshared electrons on the ________________ atom. The acid is ________________ ion. A pair of ________________ electrons on the oxygen atom of water is pushed toward the ________________ atom. Simultaneously, the pair of ________________ electrons between the hydrogen and ________________ atoms of the anilinium ion is pushed toward the ________________ atom. Thus, the oxygen– ________________ sigma bond is made and a hydrogen– ________________ sigma bond is broken. The nitrogen atom, which possessed a positive charge, is now ________________, and the oxygen atom, which was neutral, now possesses a formal ________________ charge.

________________________ ________________________

supply arrows

32. Let us consider an entire mechanism of an organic reaction. The acid-catalyzed ring opening of epoxides is initiated by protonation on the epoxide oxygen by hydronium ion.

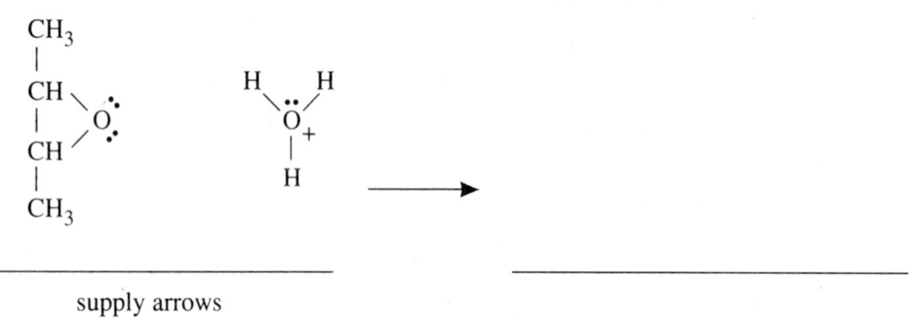

________________________ ________________________

supply arrows

33. The resulting ion then suffers nucleophilic attack on carbon by the neutral molecule, water, and the ring is opened.

________________________ ________________________

supply arrows

34. The resulting ion is then deprotonated by water giving a hydronium ion plus the organic product, a neutral 1,2-diol.

$H_2\ddot{O}:$ + $H_2O^+-CH(CH_3)-CH(CH_3)-\ddot{O}-H$ ⟶

supply arrows

Thus, all three steps in this organic reaction are examples of simultaneous sigma bond making and sigma bond breaking.

## Exercises

Here are some examples of simultaneous sigma bond making and sigma bond breaking. Where arrows are supplied, write in the products. Where the products are written in, supply the appropriate arrows.

35. $(CH_3)_2C{=}\ddot{O}$ + $H_3\ddot{O}^+$ ⟶

36. $CH_3-CH_2-\ddot{O}^+(Ph)-H$ + $:\ddot{O}(H)-H$ ⟶ $CH_3-CH_2-\ddot{O}(Ph):$ + $H_3O^+$

37. $C_6H_5-\ddot{O}:^-$ + $CH_3-CH_2-\ddot{I}:$ ⟶

38. $CH_3(-CH_2-\ddot{O})C{=}\ddot{O}$ + $H-\ddot{O}-SO_3H$ ⟶ $CH_3(CH_3-CH_2-\ddot{O})C{=}\ddot{O}^+-H$ + $:\ddot{O}^--SO_3H$

39. $\mathrm{H(Ph)C{=}\overset{+}{\ddot{O}}{-}H}$ $\quad$ $\mathrm{{}^{-}:\ddot{O}{-}C({=}\ddot{O}:){-}CH_3}$ $\longrightarrow$

______________________

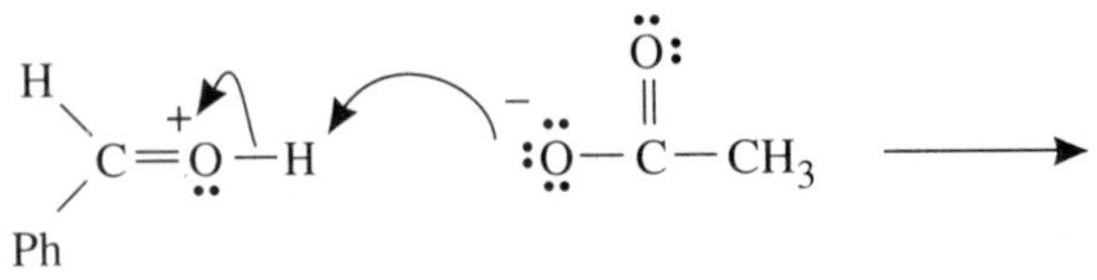

40. $\mathrm{(CH_3)_2\ddot{C}H^{-}}$ $\overset{+}{\mathrm{MgBr}}$ $\quad$ $\mathrm{CH_3{-}CH{-}CH{-}CH_3}$ (epoxide, $\ddot{O}$) $\longrightarrow$ $\mathrm{(CH_3)_2CH{-}CH(CH_3){-}CH({-}\ddot{O}:^{-}){-}CH_3}$ $\overset{+}{\mathrm{MgBr}}$

______________________

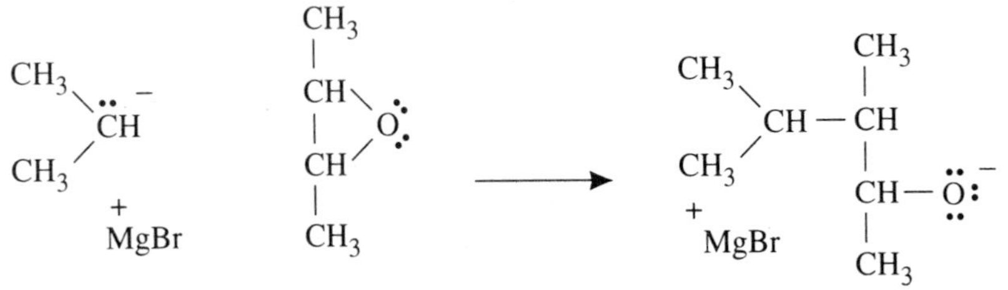

41. $\mathrm{(CH_3{-}CH_2)_2\ddot{N}{-}H}$ $\quad$ $\mathrm{H{-}\ddot{O}{-}C({=}\ddot{O}:){-}CH_3}$ $\longrightarrow$

______________________

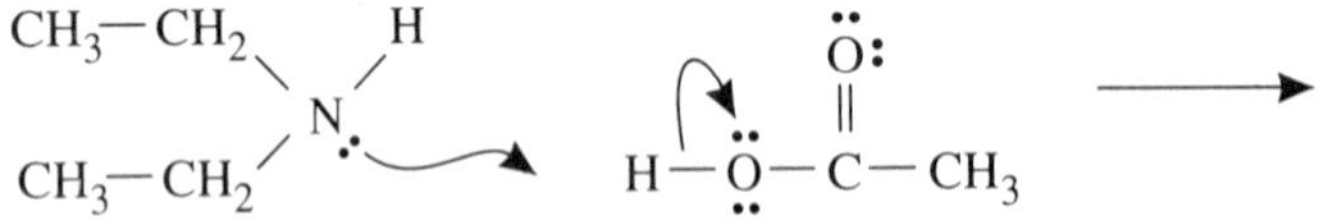

42. $\mathrm{(CH_3)_3C{-}\ddot{O}:^{-}}$ $\quad$ $\mathrm{CH_3{-}\ddot{Br}:}$ $\longrightarrow$ $\mathrm{(CH_3)_3C{-}\ddot{O}{-}CH_3}$ $\quad$ $\mathrm{:\ddot{Br}:^{-}}$

______________________

## Homolytic Sigma Bond Making and Breaking

We first saw the production of free radicals in Chapter 1. Each case involved the homolysis of a molecule in which all the electrons were paired. Uncoupling a pair of electrons requires fairly high energy. The examples in Chapter 1 used elevated temperatures or ultraviolet light. The best examples involved the scission of weak covalent bonds.

Free radical reactions occupy an important place in organic chemistry and are not so rare as the previous paragraph might suggest. Once the first few radicals are produced *(initiation),* they eagerly react with covalent bonds to produce more radicals *(propagation).* A sigma bond is broken and another is made.

A bromine radical from the photochemical homolysis of $Br_2$ reacts with one of the carbon–hydrogen bonds of methane.

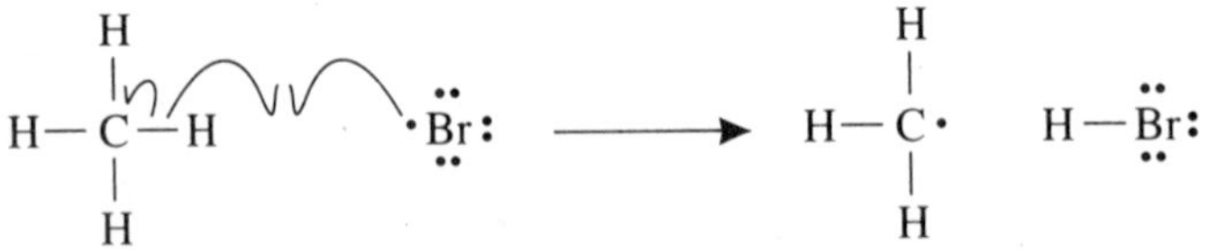

$$\mathrm{CH_4} \quad \mathrm{\cdot\ddot{Br}:} \longrightarrow \mathrm{H_3C\cdot} \quad \mathrm{H{-}\ddot{Br}:}$$

A hydrogen–bromine sigma bond is formed and a methyl radical remains.

43. A chlorine radical and ethane form HCl and an ethyl radical.

$CH_3-CH_2-H \quad \cdot\ddot{\underset{..}{Cl}}: \longrightarrow$ __________ + __________

44. A chlorine radical and cyclohexane yield the cyclohexyl radical.

(cyclohexane, H, H) $\cdot\ddot{\underset{..}{Cl}}: \longrightarrow$ __________ $+ \ H-\ddot{\underset{..}{Cl}}:$

__________

supply arrows

45. A methyl radical and a bromine molecule give bromomethane and a bromine radical.

$H_3C\cdot \quad :\ddot{\underset{..}{Br}}-\ddot{\underset{..}{Br}}: \longrightarrow$ __________ + __________

46. The hydrogen–silicon bond is susceptible to homolysis.

$Cl_3Si-H \quad \cdot\ddot{\underset{..}{O}}-C(=\ddot{O}:)-Ph \longrightarrow$ __________ $+ \ H-\ddot{\underset{..}{O}}-C(=\ddot{O}:)-Ph$

__________

supply arrows

47. Cleavage of the hydrogen–tin bond yields tin radicals.

$CH_3-\dot{C}(CN)(CH_3) \quad H-Sn-R_3 \longrightarrow$ __________ + __________

__________

supply arrows

## Exercises

Supply the arrows and the products for each of these reactions as described.

48. A chlorine radical removes a hydrogen from propane to form the isopropyl radical.

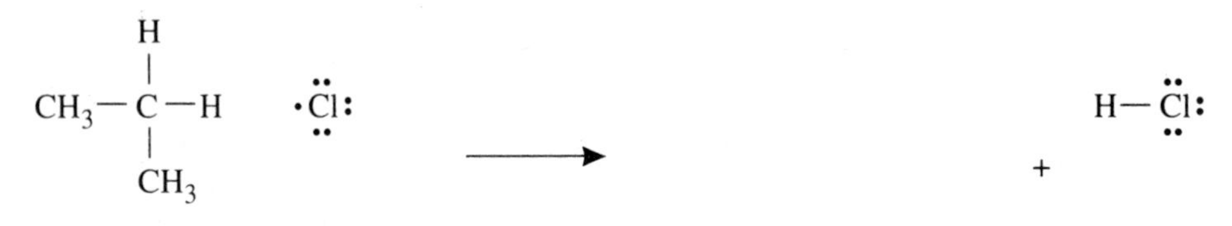


49. A chlorine radical removes a hydrogen from propene to form the allyl radical.

:Cl·   H—$CH_2$—CH=$CH_2$ ⟶ :Cl—H +

50. A benzyl radical is formed from toluene.

$C_6H_5$—$CH_2$—H   ·Br: ⟶ + H—Br:

51. Bromine is removed homolytically from 6-bromo-1-hexene.

$CH_2$=CH—$(CH_2)_3$—$CH_2$—Br:   ·$SnR_3$ ⟶ +

52. A radical center is formed on an allylic position of cyclopentene.

(cyclopentene with H, H)   ·Br: ⟶ +

53. Iodine is removed homolytically from iodocyclohexane.

(cyclohexyl)—I:   ·$SnR_3$ ⟶ +

54. The carbon radical formed in 52 removes bromine from N-bromosuccinimide.

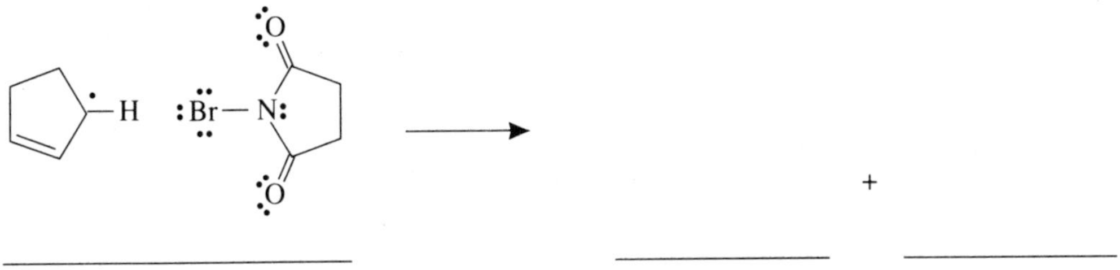

[Reaction scheme: cyclopentenyl radical (C•–H) + :Br–N: (N-bromosuccinimide) → ______ + ______]

______ ______ + ______

## Sigma Bond Making and Pi Bond Breaking

There are many steps in organic reactions in which a sigma bond is formed at the same time that a pi bond is broken. In each case, this results in a condensation of a molecule and an ion into a single ion. Actually, this category of reactions can be divided further according to whether the electrons being pushed are unshared electrons or pi electrons.

A negatively charged nucleophile may react with a pi bond as follows.

$$\ddot{Nu}^- \;\; C{=}X \longrightarrow Nu{-}C{-}\ddot{X}^-$$

In this case, the electrons being pushed are the unshared electrons on the nucleophile. The receptor is the carbon atom in the double bond, which has a pair of pushable electrons itself. Pushing the unshared electrons to the carbon atom results in the Nu–C sigma bond. Pushing the pi electrons away from the carbon atom breaks the pi bond and places a formal negative charge on the atom X. Overall, the single negative charge has been conserved.

55. The first step in the saponification of ethyl acetate involves a reaction in which hydroxide ion attacks the carbonyl-carbon atom. The nucleophile is a ________________ ion. A pair of ________________ electrons on the oxygen atom of hydroxide ion is pushed toward the carbonyl–carbon atom. The pair of *pi* electrons of the carbon–oxygen double bond of ethyl acetate is pushed toward the ________________ atom. A sigma bond has been made between oxygen and *carbon.* A pi bond has been broken between carbon and ________________. The hydroxyl oxygen, which had a negative charge, is now *neutral,* and the carbonyl oxygen, which was *neutral,* now possesses a ________________ charge.

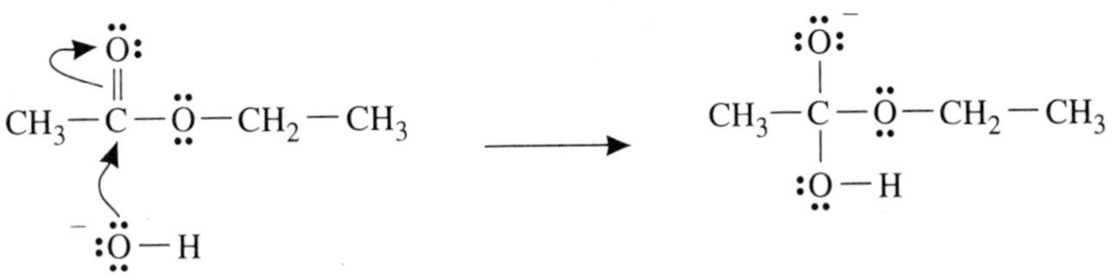

$$CH_3{-}\overset{\overset{\ddot{O}:}{\|}}{C}{-}\ddot{O}{-}CH_2{-}CH_3 \;+\; {}^-:\ddot{O}{-}H \longrightarrow CH_3{-}\overset{:\ddot{O}:^-}{\underset{:\ddot{O}-H}{C}}{-}\ddot{O}{-}CH_2{-}CH_3$$

56. Similarly, amides are attacked by hydroxide ion. The nucleophile is ________________ ion. A pair of ________________ electrons on the hydroxyl oxygen atom is pushed toward the carbonyl-carbon atom. The pair of ____________ electrons in the carbon- ______________ double bond is pushed toward the ________________ atom. A sigma bond has been made between oxygen and ________________. A pi bond has been broken between carbon and ________________. The hydroxyl oxygen, which had a negative charge, is now ________________. The carbonyl oxygen, which was neutral, now possesses a ________________ charge.

$C_6H_5-C(=\ddot{O}:)-\ddot{N}H-CH_3$

$^{-}:\ddot{O}-H$ ⟶

____________________ ____________________

supply arrows

57. The first step in the formation of a cyanohydrin under neutral or basic conditions involves a similar process.

$\ddot{N}\equiv\ddot{C}^{-}$ $\quad R_2C=\ddot{O}:$ ⟶

____________________ ____________________

supply arrows

58. Since the alkyl portion of a Grignard reagent can be treated as a carbanion, its well known reaction with a carbonyl group fits this category.

$CH_3-CH_2-\ddot{C}H_2^{-}$

$^{+}MgBr$

$H-C(=\ddot{O}:)-CH_2-CH(CH_3)-CH_3$ ⟶

____________________ ____________________

supply arrows

59. The pi bond that is broken can be part of a triple bond. Consider the first step in the base-catalyzed hydrolysis of a nitrile.

$$C_6H_5\text{—}C{\equiv}N: \qquad {}^{-}:\ddot{\underset{..}{O}}\text{—}H \longrightarrow \underline{\qquad\qquad}$$

______________________ supply arrows ______________________

*　　*　　*　　*　　*

A slight variation of this process occurs when the nucleophile possessing the unshared electrons is neutral. In virtually all of these cases, the pi systems must have suffered some kind of previous activation in order for the reaction to proceed at a reasonable rate. The activation process (usually protonation) puts a positive charge on the pi system.

$$\ddot{Nu} \quad \rangle C{=}X^{+} \longrightarrow \overset{+}{Nu}\text{—}\underset{|}{\overset{|}{C}}\text{—}\ddot{X}$$

60. We have already seen that hydroxide ion, a strong nucleophile, will attack the carbonyl – carbon atom of an ester. Analogously, water, a weak nucleophile, will attack the carbonyl–carbon atom of a protonated ester. A pair of unshared electrons on the oxygen atom of the water molecule is pushed toward the carbonyl–carbon atom. The pair of *pi* electrons of the carbon–oxygen double bond is pushed toward the *oxygen* atom. Thus, a sigma bond has been made between oxygen and ________________, and a pi bond has been broken between carbon and ________________. The oxygen atom of the water molecule, which was neutral, now possesses a ________________ charge. The carbonyl–oxygen atom, which had a positive charge, is now ________________.

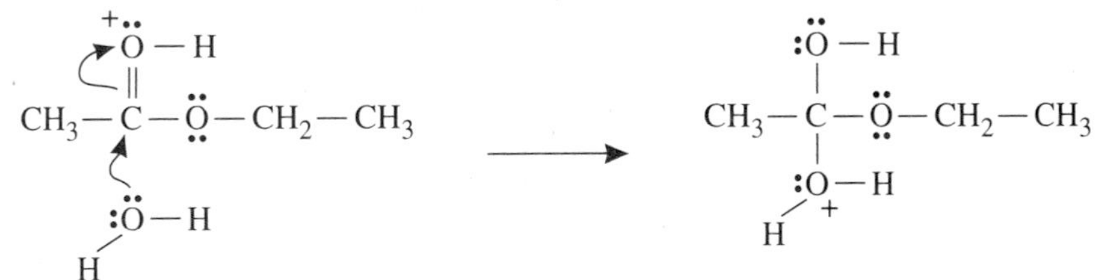

61. An important step in the production of acetals occurs when a molecule of alcohol undergoes reaction with the protonated carbonyl group of an aldehyde. A pair of ________________ electrons on the ________________ atom of the alcohol is pushed toward the carbonyl–carbon atom. The pair of ________________ electrons of the carbon–oxygen double bond is pushed toward the ________________ atom. Thus, a sigma bond is made between oxygen and ________________, and a pi bond has been broken between carbon and ________________.

The oxygen atom of the alcohol, which was neutral, now possesses a _______________ charge. The oxygen atom of the protonated aldehyde, which had a positive charge, is now _______________.

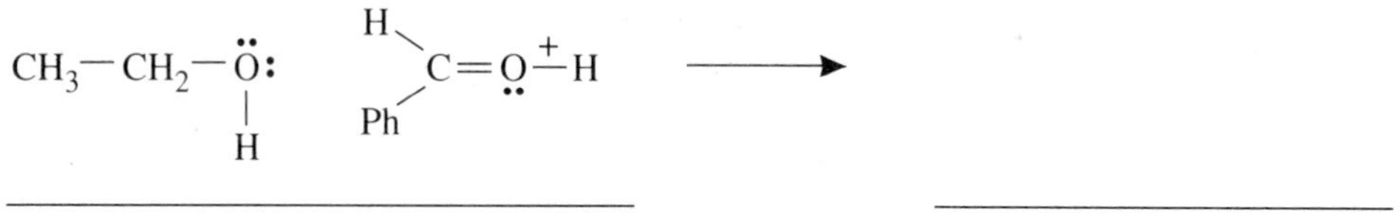

$$CH_3-CH_2-\ddot{O}(H): \quad + \quad H(Ph)C=\overset{+}{\ddot{O}}-H \longrightarrow$$

_______________ supply arrows _______________

62. The acid-catalyzed hydrolysis of nitriles involves this kind of reaction on a triple bond.

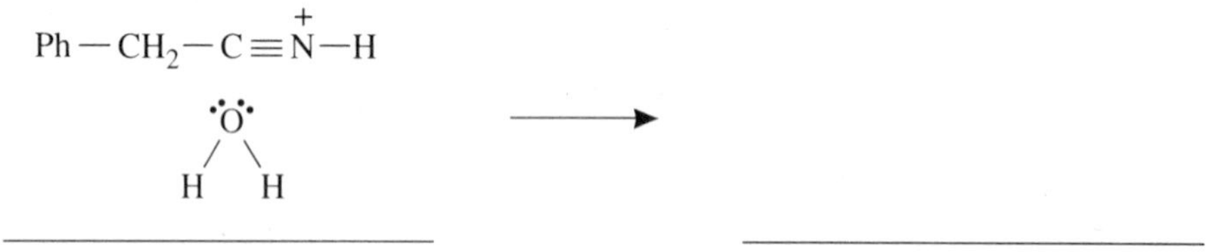

$$Ph-CH_2-C\equiv\overset{+}{N}-H \quad + \quad H_2\ddot{O}: \longrightarrow$$

_______________ supply arrows _______________

* * * * *

We have seen a number of examples of the simultaneous sigma bond making and pi bond breaking type of reaction where the pushable electrons were unshared electrons. In many mechanisms, one encounters steps in which there is simultaneous sigma bond making and pi bond breaking and it is the pi electrons that are pushed.

$$\overset{+}{E} \quad -\underset{|}{C}=Z \longrightarrow E-\underset{|}{\overset{|}{C}}-\overset{+}{Z}$$

Once again, an ion and a molecule condense into a single ion. The electrons being pushed are the pi electrons of an unsaturated system. The receptor is an electrophile, E, usually possessing a formal positive charge. Pushing the pi electrons to the electrophile results in the E–C sigma bond and breaks the pi bond. A formal positive charge is left on the atom Z, thus conserving the charge in the system.

In a typical Friedel–Crafts alkylation of benzene, a *t*-butyl cation reacts with a molecule of benzene. A pair of *pi* electrons from the aromatic ring is pushed toward the positively charged *carbon* atom of the *t*-butyl cation. Thus, a pi bond between carbon and *carbon* is broken, and a sigma bond between carbon and *carbon* is made. The positively charged carbon atom of the *t*-butyl cation is now *neutral.* The carbon atom, *from which the pi electrons departed,* now possesses a *positive* charge.

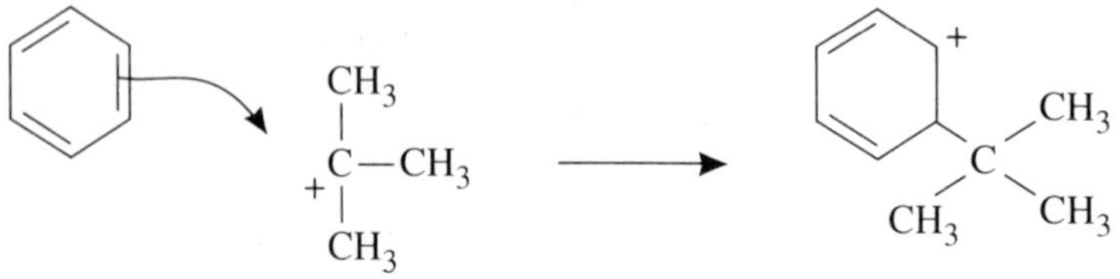

$$C_6H_6 \quad + \quad \overset{+}{C}(CH_3)_3 \longrightarrow \text{(cyclohexadienyl cation)}-C(CH_3)_3$$

Here we see why it is necessary to preserve the Kekulé structure for benzene (Chapter 2). For clarity it is expedient to push electrons in pairs, and that is how it is done above. The reaction, as it actually occurs, is quite a complicated affair involving the entire pi system of the benzene ring.

63. Consider the alkylation of *p*-xylene by an isopropyl cation. A pair of ____________ electrons from the aromatic ring is pushed toward the positively charged carbon atom of the isopropyl cation. Thus, a pi bond is broken between carbon and ________________, and a sigma bond is made between carbon and ________________ The positively charged carbon atom of the isopropyl cation is now ________________ The carbon atom, from which the pi electrons departed, now possesses a ________________ charge.

$CH_3$ $CH_3$ (p-xylene) + $\overset{+}{C}H(CH_3)-CH_3$ ⟶

________________ ________________

supply arrow

64. The condensation step in the dimerization of 2-methylpropene is very similar. The pi electrons of the double bond in a 2-methylpropene molecule are pushed toward the *t*-butyl cation. However, there are two ways in which this can be done. One way results in a tertiary cation.

$CH_3-C(CH_3)=CH_2$ + $\overset{+}{C}(CH_3)_2-CH_3$ ⟶ $CH_3-\overset{+}{C}(CH_3)-CH_2-C(CH_3)_2-CH_3$

________________ ________________

supply arrow

Another way results in a primary cation.

$CH_3-C(CH_3)=CH_2$ + $\overset{+}{C}(CH_3)_2-CH_3$ ⟶

________________ ________________

supply arrow

The stabilities of the resulting carbocation (actually, the stabilities of the transition states) determine that the former process predominates.

65. It is convenient to view protonation of a pi system as occurring by this simple process. Thus, a pair of ________________ electrons from 2-butene is pushed toward the proton. A carbon–________________ pi bond has been broken, and a carbon–________________ sigma bond has been made. The proton, which had a positive charge, is now ________________. The carbon atom from which the pi electrons departed now possesses a ________________ charge.

$CH_3-CH=CH-CH_3$ $\longrightarrow$

$H^+$

________________ ________________

supply arrow

66. The protonation of 1-methylcyclohexene occurs in a similar fashion. Here you must make the proper choice between making a bond between the proton and $C_1$ or the proton and $C_2$.

1-methylcyclohexene (C1 bearing $CH_3$, C2 bearing H) + $H^+$ $\longrightarrow$

________________ ________________

supply arrow

However, as will be shown in a later section, protonation is usually more complicated since it is inaccurate to propose an uncombined proton as a viable species.

## Pi Bond Making and Sigma Bond Breaking

There are many steps in organic reaction mechanisms where a pi bond is formed at the same time that a sigma bond is broken. Each of the reactions in the preceding section, when written in reverse, provides an example of this process.

$$Nu-\overset{|}{\underset{|}{C}}-\ddot{Z}^- \longrightarrow \ddot{Nu}^- \quad >C=Z$$

The pushable electrons are the unshared electrons on the atom Z. As they are pushed toward the carbon atom to form a pi bond, the sigma electrons that form the Nu–C bond leave with Nu. Charge is conserved as Nu leaves as an anion.

Consider the fate of the ion formed from the attack of hydroxide ion on ethyl acetate. If one of the pairs of unshared electrons on the negatively charged oxygen atom is pushed toward carbon to reform the carbonyl pi bond and the sigma bond between carbon and the OH group is broken, we have exactly reversed the reaction by which the ion was formed.

$$CH_3-\overset{\overset{:\ddot{O}:^-}{|}}{\underset{\underset{:\ddot{O}-H}{|}}{C}}-\ddot{O}-CH_2-CH_3 \longrightarrow CH_3-\overset{\overset{\ddot{O}:}{\|}}{C}-\ddot{O}-CH_2-CH_3 \quad ^-:\ddot{O}-H$$

This step occurs. However, since it does not lead toward the products of saponification, namely, carboxylate anion and alcohol, we shall consider another fate for the ion.

67. If, as the unshared electrons are being used to reform the carbonyl pi bond, the sigma bond between the carbon and the $-\ddot{O}-CH_2-CH_3$group is broken, we get something that looks like progress toward products. Thus, a pi bond has been made between oxygen and __________________, and a sigma bond has been broken between carbon and __________________.

$$CH_3-\overset{\overset{:\ddot{O}:^-}{|}}{\underset{\underset{:\ddot{O}-H}{|}}{C}}-\ddot{O}-CH_2-CH_3 \longrightarrow$$

__________________________ ____________________

supply arrows

The simple transfer of a proton from the carboxylic acid to the alcoholate anion leads to the saponification products.

68. The final step in the benzoin condensation involves the making of a pi bond between oxygen and carbon and the breaking of the sigma bond between carbon and the CN group.

$$C_6H_5-\overset{\overset{CN}{|}}{\underset{\underset{:\ddot{O}:^-}{|}}{C}}-\underset{\underset{:\ddot{O}-H}{|}}{CH}-C_6H_5 \longrightarrow$$

__________________________ ____________________

supply arrows

*     *     *     *     *

Another general form that these steps can take involves the expulsion of a neutral nucleophile.

$$\overset{+}{Nu}-\overset{|}{\underset{|}{C}}-\ddot{X} \longrightarrow \ddot{Nu} \quad >C=X^{+}$$

69. The cleavage step in the acid-catalyzed hydrolysis of esters provides a good example. A pair of unshared electrons from one of the OH groups is pushed toward the *carbon* atom. Simultaneously, the pair of *sigma* electrons is pushed toward the positively charged oxygen atom. Thus, a pi bond is made between oxygen and ________________, and a sigma bond is broken between carbon and ________________. The neutral oxygen atom from which the pi electrons were pushed now possesses a ________________ charge. The positively charged oxygen toward which the electrons were pushed is now ________________.

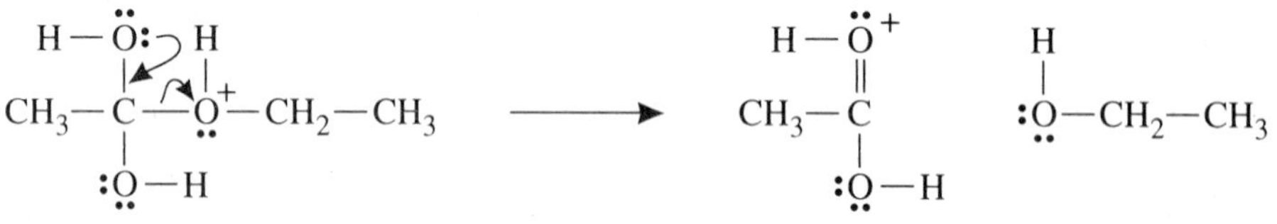

$$CH_3-\overset{H-\ddot{O}:}{\underset{:\ddot{O}-H}{C}}-\overset{H}{\overset{|}{\ddot{O}^{+}}}-CH_2-CH_3 \longrightarrow CH_3-\overset{H-\ddot{O}^{+}}{\overset{\|}{\underset{:\ddot{O}-H}{C}}} \quad \overset{H}{\overset{|}{:\ddot{O}}}-CH_2-CH_3$$

Here, the resulting conjugate acid of acetic acid loses a proton to complete the hydrolytic mechanism.

70. Similarly, the penultimate step in the hydrolysis of an acetal is the loss of a methanol molecule.

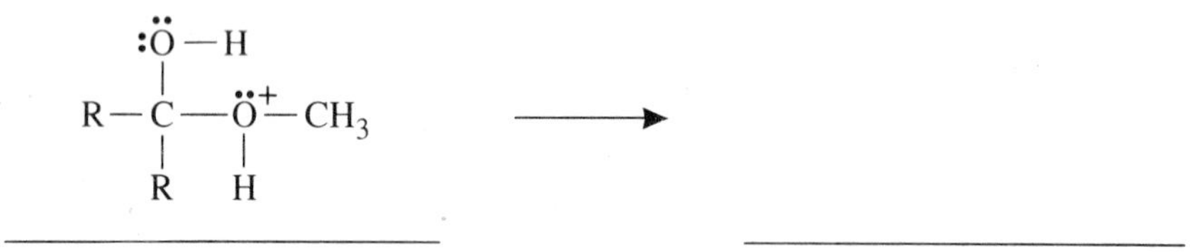

$$R-\overset{:\ddot{O}-H}{\underset{R}{C}}-\underset{H}{\ddot{O}^{+}}-CH_3 \longrightarrow$$

________________ ________________

supply arrows

* * * * *

The examples of simultaneous pi bond formation and sigma bond breaking that we have seen so far have occurred when a pair of unshared electrons was pushed to form the pi bonds, and a nucleophile, either neutral or negatively charged, was expelled. Another way in which a pi bond can be formed when a sigma bond is broken is to push a pair of sigma electrons in such a way as to form the pi bond. In these cases a positively charged fragment is lost.

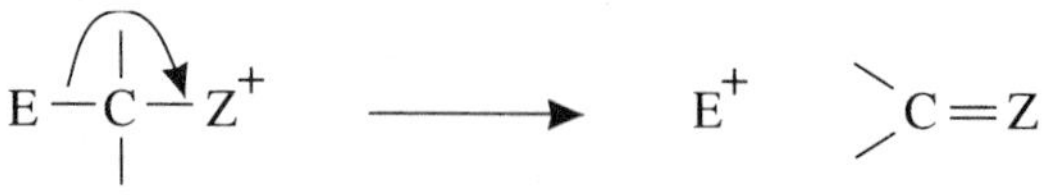

$$E-\overset{|}{\underset{|}{C}}-Z^{+} \longrightarrow E^{+} \quad >C=Z$$

Most commonly, this occurs when a proton is lost from some positively charged intermediate.

71. The *t*-butyl cation can lose a proton from a carbon next to the carbocation and form the neutral molecule, isobutylene. Thus, a pair of sigma electrons forming a *carbon–hydrogen* bond is pushed toward the positively charged *carbon* atom. The sigma bond between hydrogen and ________________ is broken, and a pi bond is formed between carbon and ________________. The neutral hydrogen atom now possesses a ________________ charge. The positively charged carbon atom is now ________________.

$$H_2C{-}H \text{ on } CH_3{-}\overset{+}{C}{-}CH_3 \longrightarrow H_2C{=}C(CH_3){-}CH_3 \quad H^+$$

72. Similarly, a proton can be lost from *sec*-butyl cation to form 2-butene. A pair of sigma electrons which form the hydrogen-carbon bond is pushed toward the positively charged ________________ atom. The sigma bond between hydrogen and ________________ is broken, and a pi bond between carbon and ________________ is made. The neutral hydrogen atom now possesses a ________________ charge. The positively charged carbon

$$CH_3{-}\underset{|}{\underset{H}{CH}}{-}\overset{+}{CH}{-}CH_3 \longrightarrow$$

____________________ ____________________

supply arrow

73. Consider the loss of a proton from the carbocation below. This intermediate could lose a primary hydrogen giving the terminal olefin, 2,4,4-trimethyl-1-pentene.

$$\underset{|}{\underset{H}{CH_2}}{-}\overset{CH_3}{\overset{|}{\underset{+}{C}}}{-}CH_2{-}\overset{CH_3}{\overset{|}{\underset{\underset{CH_3}{|}}{C}}}{-}CH_3 \longrightarrow$$

____________________ ____________________

supply arrow

Conversely, it could lose a secondary hydrogen and become 2,4,4-trimethyl-2-pentene.

$$CH_3{-}\overset{CH_3}{\overset{|}{\underset{+}{C}}}{-}\underset{\underset{H}{|}}{CH}{-}\overset{CH_3}{\overset{|}{\underset{\underset{CH_3}{|}}{C}}}{-}CH_3 \longrightarrow$$

____________________ ____________________

supply arrow

The latter process predominates since the olefin formed is more stable. The reformation of the aromatic system in the last step of electrophilic aromatic-substitution involves this process.

74. In the final step of the chlorination of benzene, the positively charged intermediate loses a proton from the carbon atom to which a chlorine atom has been attached. The pair of sigma electrons that form the bond between *carbon* and *hydrogen* is pushed toward the positively charged carbon atom. The result is the breaking of a hydrogen–______________ sigma bond and the making of a carbon–______________ pi bond. (The 6-pi electron aromatic system is restored.) The neutral hydrogen atom now possesses a ______________ charge and the positively charged carbon atom is now______________.

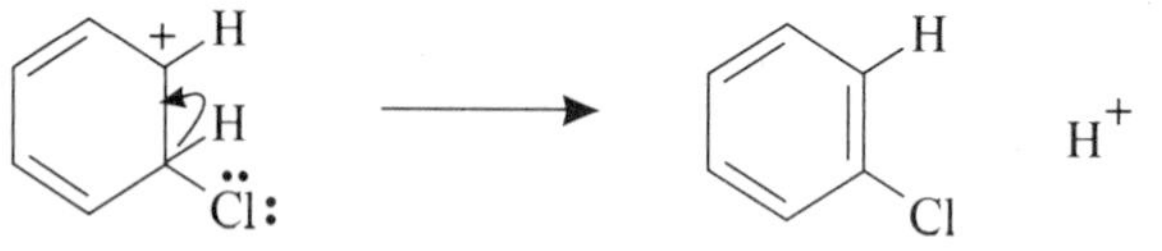

75. The final step in the nitration of toluene is similar.

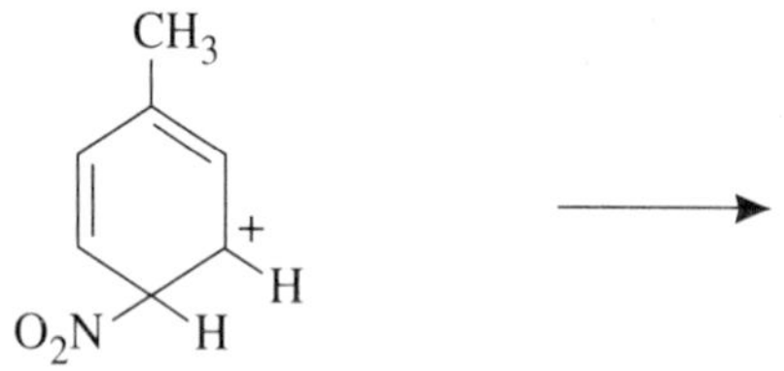

______________ ______________

supply arrow

76. Also, the final step in the bromination of *m*-xylene.

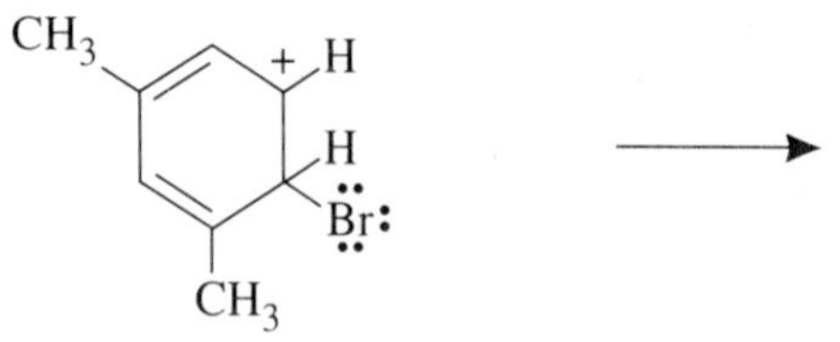

______________ ______________

supply arrow

77. On rare occasions a *t*-butyl group is lost in electrophilic aromatic substitution.

$CH_3$ $CH_3$ H C $CH_3$ + H Ph C $CH_3$ :Ö $\longrightarrow$ __________ + $CH_3$ $CH_3$ $\overset{+}{C}$ $CH_3$

__________

supply arrow

78. Finally, the loss of a proton from the conjugate acid of a carbonyl group is often found as the final step in a reaction mechanism.

H + H $\overset{+}{C}$—Ö: H $\longrightarrow$ H C=Ö: H $\quad H^+$

__________ __________

supply arrow

79. Similarly,

H | Ö: $CH_3$—$\overset{+}{C}$ Ö: | H $\longrightarrow$

__________ __________

supply arrow

## Exercises

Where arrows are supplied write in the products; where the products are written supply the appropriate arrows.

80. O O —$CH{=}CH{-}NO_2$ $\longrightarrow$ O O —CH—$\ddot{C}H^-$—$NO_2$ | :Ö—H

$^-$:Ö—H

__________

81.

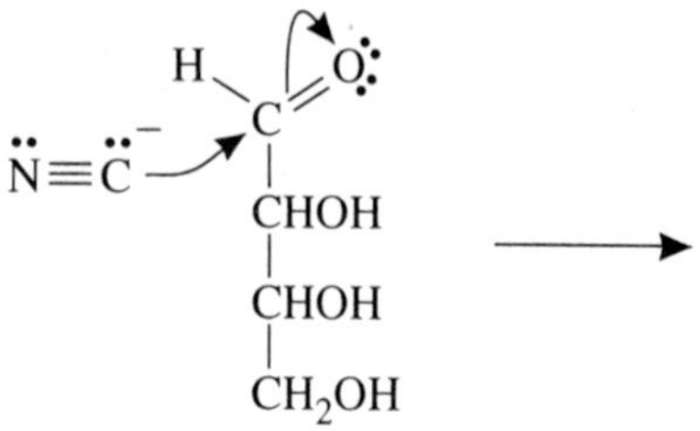

$$:N\equiv \ddot{C}:^- \quad \text{H—C(=\ddot{O}:)—CHOH—CHOH—CH}_2\text{OH} \longrightarrow$$

82.

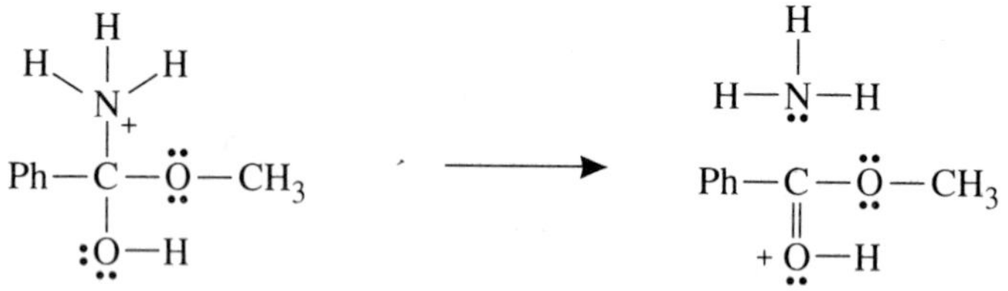

$$\text{Ph—C(—N}^+\text{H}_3\text{)(—\ddot{O}—CH}_3\text{)—\ddot{O}—H} \longrightarrow \text{H—\ddot{N}(H)—H} \quad \text{Ph—C(—\ddot{O}—CH}_3\text{)=}\overset{+}{\ddot{O}}\text{—H}$$

83.

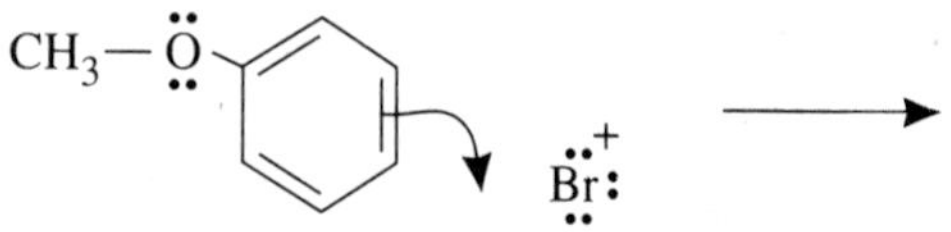

$$CH_3—\ddot{O}—C_6H_5 \quad :\ddot{Br}:^+ \longrightarrow$$

84.

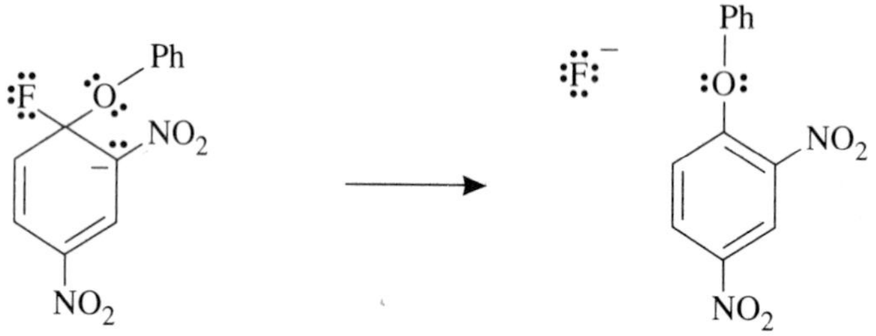

:F: O—Ph, $NO_2$, $NO_2$ ⟶ $:\ddot{F}:^-$ Ph—:O:, $NO_2$, $NO_2$

85.

$H_2\ddot{N}$—cyclohexyl, :Cl:, $O_2N$, $NO_2$, $NO_2$ ⟶

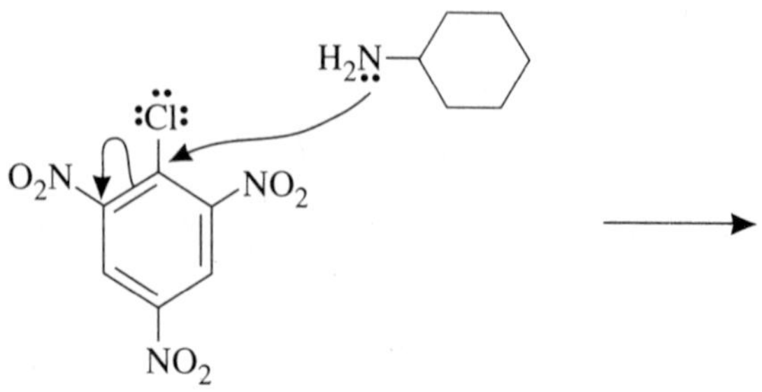

86. CH—CH$_2$—NO$_2$ :O:$^-$ ⟶ CH =O: $^-$CH$_2$—NO$_2$

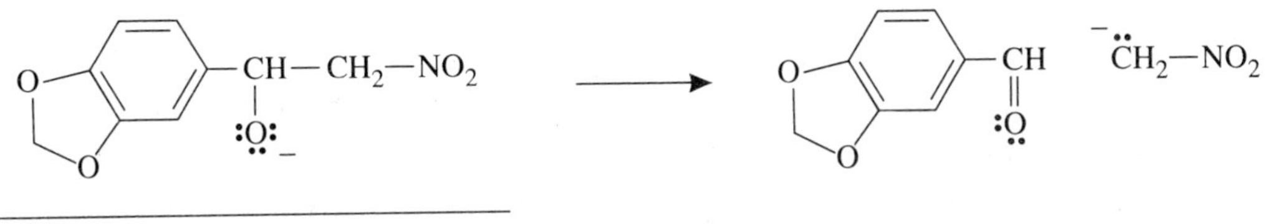

87. ⟶

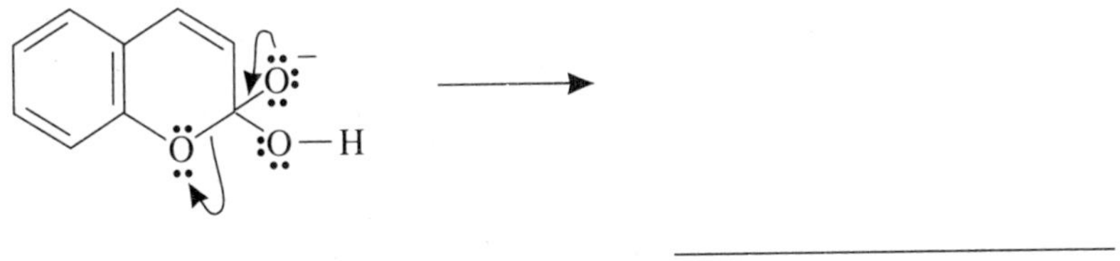

88. $^+$NH$_2$ Ph—C—O—CH$_3$ :O—H H ⟶ :NH$_2$ Ph—C—O—CH$_3$ :O$^+$—H H

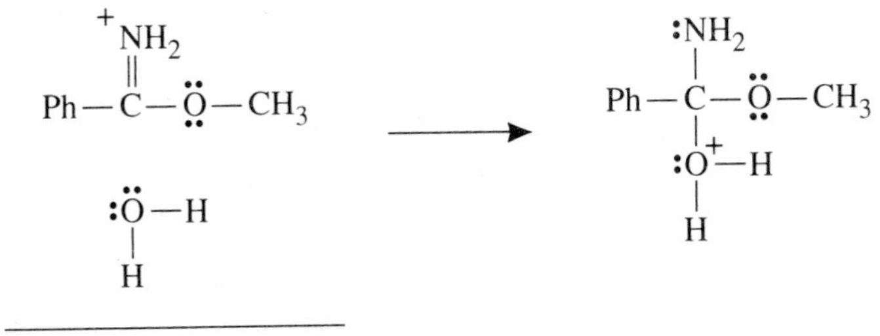

89. H O$^+$ :O—CH$_3$ H ⟶

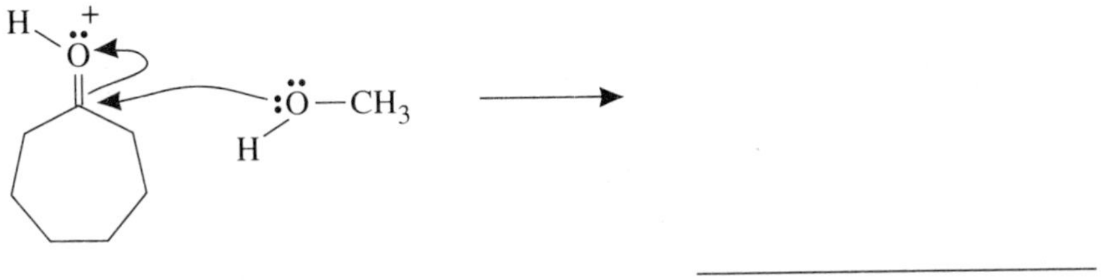

90. O: C C + O: ⟶ O: C C + O:

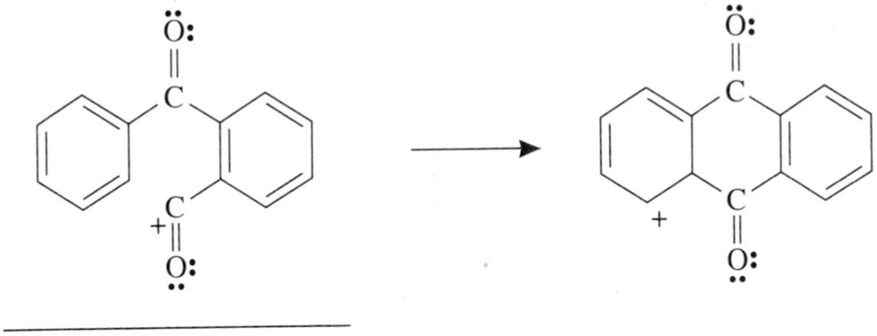

91. :F: O$_2$N NO$_2$ $^-$:O ⟶

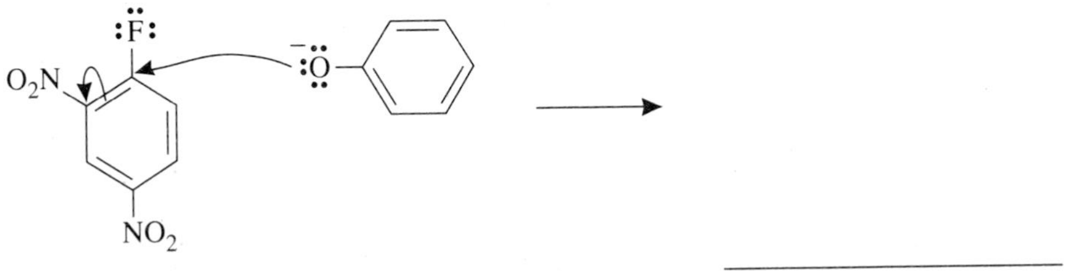

92. 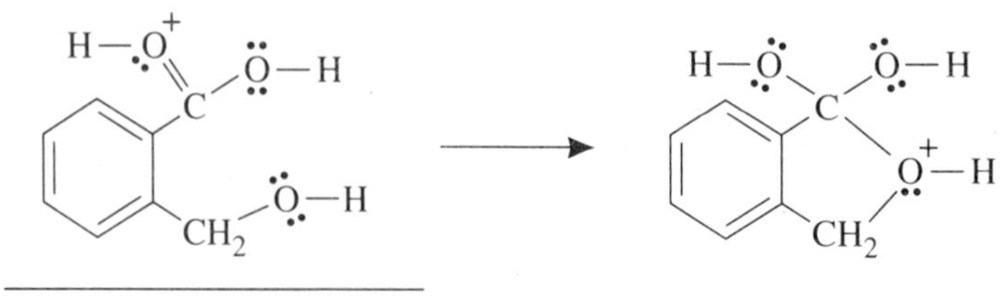 $H-\ddot{O}^+=C(-\ddot{O}-H)$ (benzene ring with $CH_2-\ddot{O}-H$) ⟶ $H-\ddot{O}-C(-\ddot{O}-H)$ (benzene ring, $CH_2$, $\overset{+}{O}-H$ ring)

______________________

## Homolytic Sigma Bond Making and Pi Bond Breaking

Free radicals will react with pi systems. Such reactions usually result in addition to double bonds.

Hydrogen bromide will add to 2-butene in a free radical reaction. Initiation provides bromine radicals.

$$:\ddot{Br}-H \quad \cdot\ddot{O}-\overset{\overset{\ddot{O}:}{\|}}{C}-Ph \longrightarrow :\ddot{Br}\cdot + H-\ddot{O}-\overset{\overset{\ddot{O}:}{\|}}{C}-Ph$$

The bromine radical takes one of the pi electrons from 2-butene, leaving an unpaired electron on carbon.

$$CH_3-CH=CH-CH_3 \quad :\ddot{Br}\cdot \longrightarrow CH_3-\underset{\underset{:\ddot{Br}:}{|}}{CH}-\dot{C}H-CH_3$$

The carbon radical abstracts a hydrogen from H–Br, giving the addition product, 2-bromobutane, plus another bromine radical that can propagate the reaction.

$$CH_3-\underset{\underset{:\ddot{Br}:}{|}}{CH}-\dot{C}H-CH_3 \quad H-\ddot{Br}: \longrightarrow CH_3-\underset{\underset{:\ddot{Br}:}{|}}{CH}-\underset{\underset{H}{|}}{CH}-CH_3 \quad \cdot\ddot{Br}:$$

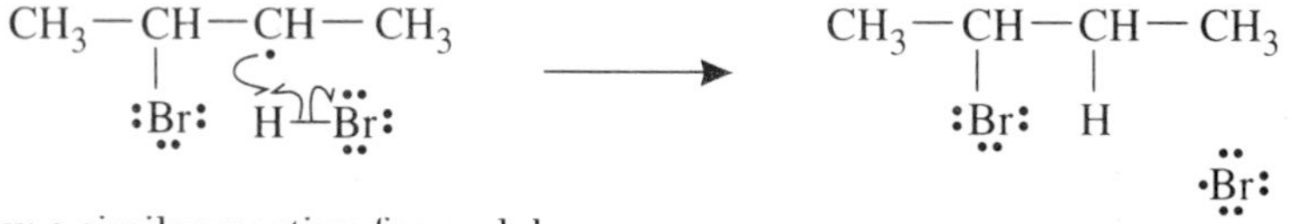

93. (a) Show a similar reaction for cyclohexene.

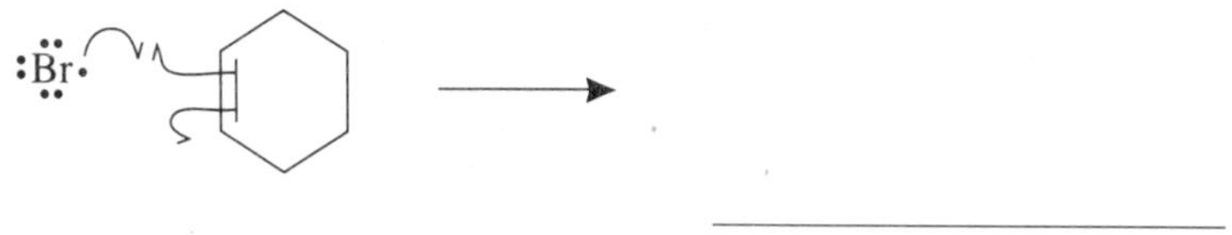

$$:\ddot{Br}\cdot \quad \text{cyclohexene} \longrightarrow$$

______________________

(b) Show the reaction of the carbon radical with H–Br to propagate the reaction.

$$\text{(2-bromocyclohexyl radical)} \quad :\ddot{Br}-H \longrightarrow :\ddot{Br}\cdot +$$

______________________

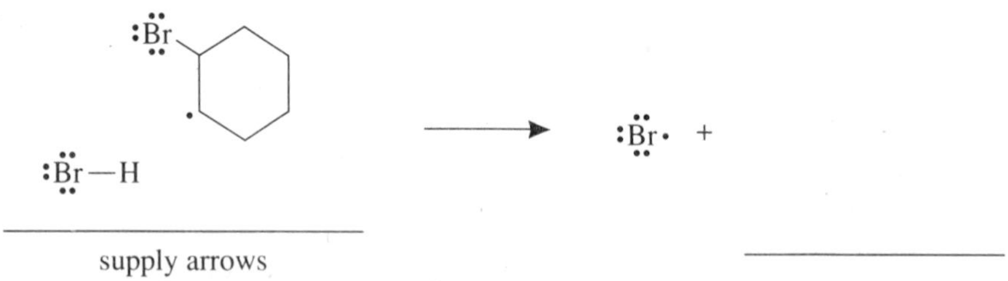

supply arrows

______________________

When the alkene is unsymmetrical, the reaction takes one of two possible pathways.

94. (a) The radical attack might lead to the primary carbon radical.

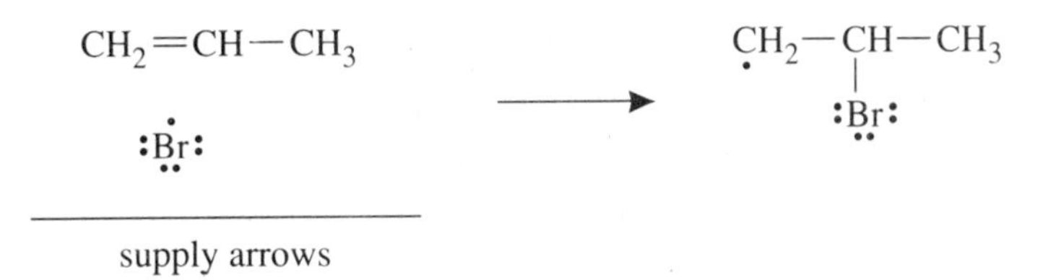


(b) Actually, attack leads to the more stable secondary radical.

$CH_2{=}CH{-}CH_3$

$:\dot{Br}:$ ⟶ ____________

____________

supply arrows

Other radicals add to double bonds.

95.

$Cl_3C\cdot$ $CH_2{=}CH{-}CH_2{-}CH_2{-}CH_3$ ⟶ ____________

96.

$Cl_3Si\cdot$ $CH_2{=}CH{-}CH(CH_3)_2$ ⟶ ____________

____________

supply arrows

97. Carbon radicals will add to carbon–carbon double bonds.

(cyclopentyl radical)· $CH_2{=}CH{-}CN$ ⟶ ____________

____________

supply arrows

98. (a) A radical will attack carbon–sulfur double bonds.

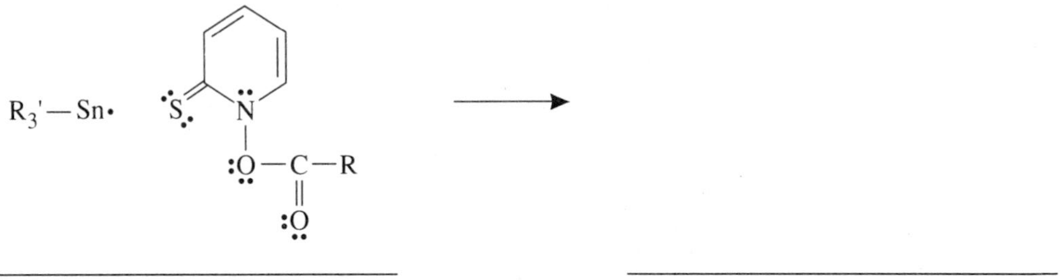

$R_3'$—Sn•   :S: N   :O—C—R   ‖   :O   ⟶

supply arrows

(b) Actually, part (a) is the first step in a clever way to produce R•. See if you can push electrons to produce the products.

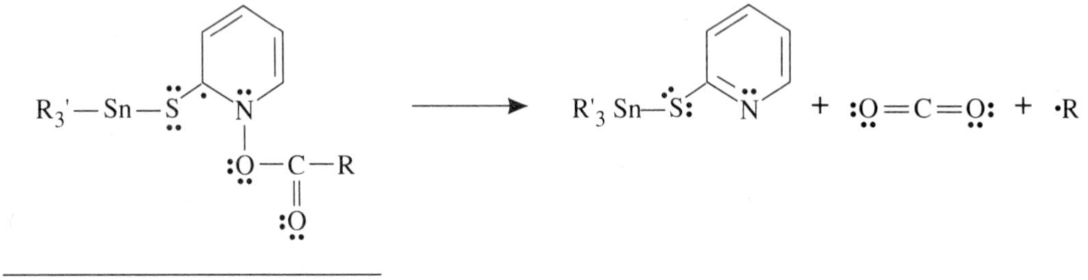

$R_3'$—Sn—S N   :O—C—R   ‖   :O   ⟶   $R'_3$ Sn—S: N + :O=C=O: + •R

supply arrows

## Complex Mechanisms

When pushing electrons in this chapter, we have, until now, illustrated steps in reaction mechanisms with examples in which the making and breaking of no more than two bonds occurs in any step. The awful truth is that there are many examples that are more complex. We shall take up some of them in this section. It should be emphasized that these steps are more complicated in numbers of electron pairs being pushed, but the types of bond making and bond breaking should all be familiar by this time.

The E2 elimination is a reaction in which everything happens at once. In the example below a base abstracts a proton *beta* to the leaving group, X. The electrons that had formed the C–H bond are pushed toward the *alpha* carbon from which, in turn, the leaving group departs, taking with it the pair of electrons that had formed the C–X bond. All of this happens more or less simultaneously.

Inspection reveals that this process involves sigma bond making, *sigma* bond breaking, pi bond making and sigma bond breaking.

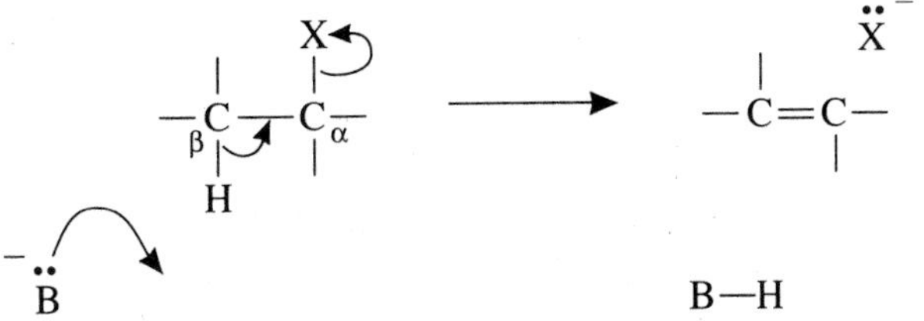

99. In the formation of 2-butene from 2-bromobutane a pair of unshared electrons from the oxygen atom of hydroxide ion is pushed toward the beta hydrogen forming an oxygen–*hydrogen* bond. The pair of sigma electrons which was the hydrogen–carbon bond is pushed toward the alpha carbon atom forming a pi bond between carbon and *carbon.* The sigma electrons of the carbon–bromine bond are pushed toward the *bromine* atom. The hydroxyl oxygen, which had a negative charge, is now *neutral* and the bromine atom, which was neutral, now has a *negative* charge.

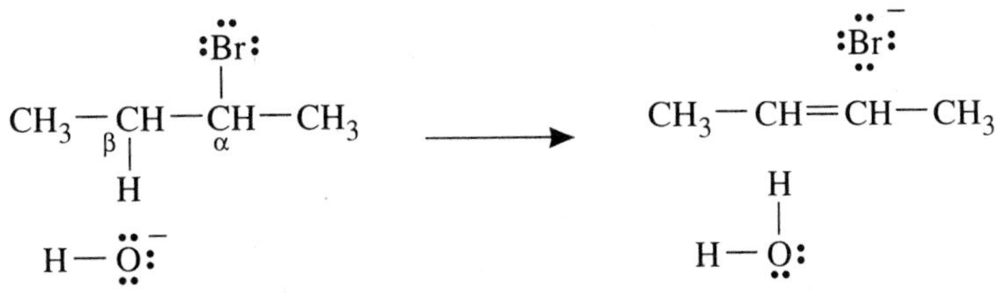

$$CH_3-\underset{\beta}{CH}(H)-\underset{\alpha}{CH}(Br)-CH_3 + H-\ddot{O}:^- \longrightarrow CH_3-CH=CH-CH_3 + H-O-H + :\ddot{Br}:^-$$

supply arrows

100. In the reaction of ethoxide ion with *t*-butyl chloride, a pair of unshared electrons on the oxygen atom of ethoxide ion is pushed toward the beta hydrogen forming an oxygen–____________ bond. The pair of sigma electrons that was the hydrogen–carbon bond is pushed toward the alpha carbon forming a pi bond between carbon and ____________. The sigma electrons of the carbon–chlorine bond are pushed toward the chlorine atom. The ethoxyl oxygen, which had a negative charge, is now ____________. The chlorine, which was neutral, now has a ____________ charge.

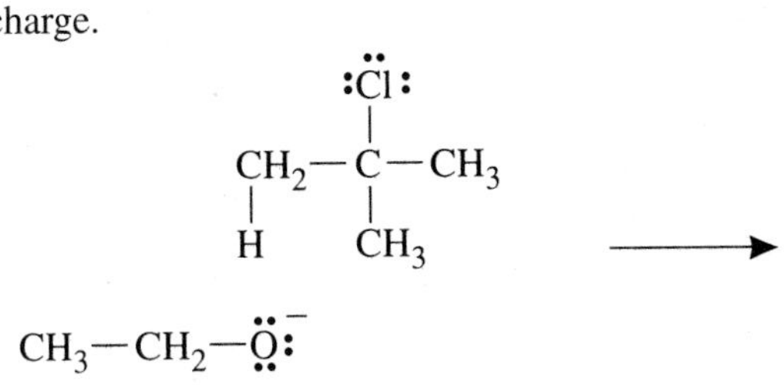

$$CH_2(H)-C(Cl)(CH_3)-CH_3 + CH_3-CH_2-\ddot{O}:^- \longrightarrow$$

supply arrows

101. Here is an example of an E2 reaction forming a double bond in a ring.

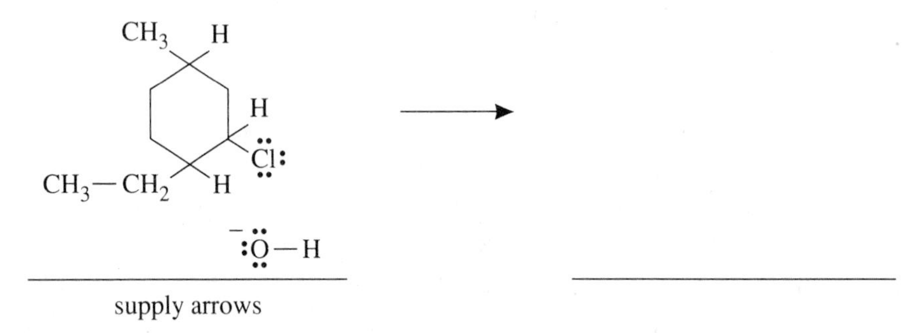


102. The Hofmann elimination is a similar reaction except that the charge distribution is different. The hydroxyl oxygen, which had a negative charge, becomes __________________, and the nitrogen, which had a positive charge, becomes __________________.

$$\overset{+}{N}(CH_3)_3$$

$$CH_2-CH-CH_2-CH_3$$

$$H$$

$$H-\ddot{\underset{..}{O}}:^-$$

$\longrightarrow$

________________________ ________________________

supply arrows

103. Here is a Hofmann elimination in which a double bond is formed in the ring.

$$H \quad \overset{+}{N}(CH_3)_3$$

$$H \quad H$$

$$^-:\ddot{\underset{..}{O}}-CH_2-CH_3$$

$\longrightarrow$

________________________ ________________________

supply arrows

104. Finally, the Hofmann elimination can result in a ring-opening reaction.

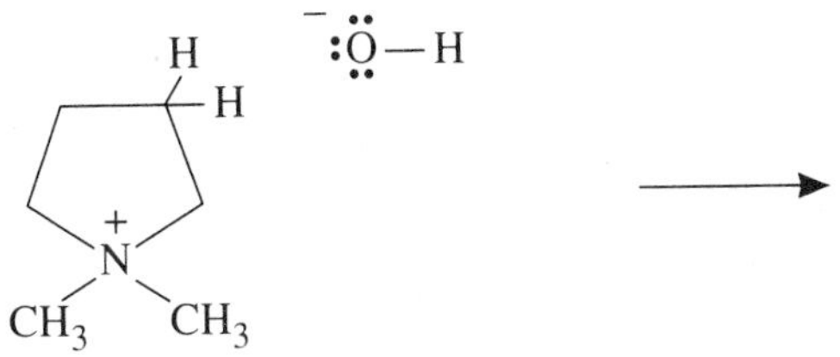

$\longrightarrow$

supply arrows

105. When a chlorine molecule interacts with a carbon–carbon double bond, the result is a **carbocation.** The pi electrons of the carbon–carbon double bond are mobilized to form a carbon–chlorine sigma bond. In turn, the sigma electrons of the chlorine–chlorine bond are pushed so that the chloride ion departs. This process involves pi bond breaking, sigma bond making, and sigma bond breaking.

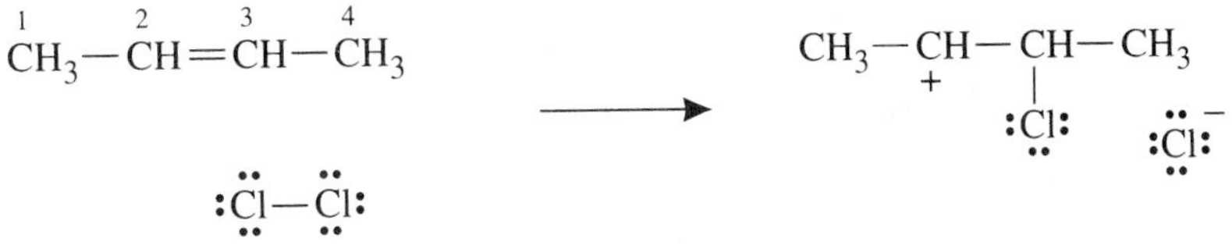

$\overset{1}{C}H_3-\overset{2}{C}H=\overset{3}{C}H-\overset{4}{C}H_3 \quad :Cl-Cl: \longrightarrow CH_3-\overset{+}{C}H-CH(Cl)-CH_3 \quad :Cl:^-$

supply arrows

The carbon atom, $C_2$, from which the pi electrons depart obtains a ____________ charge, and the chlorine atom which the electrons are pushed obtains a ____________ charge.

106. An unionized acid protonates a carbon–carbon double bond in an analogous fashion.

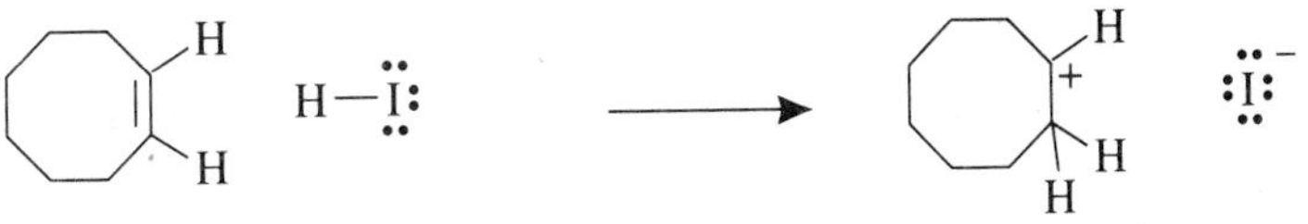

$H-I: \longrightarrow$ $:I:^-$

supply arrows

107. Another example is

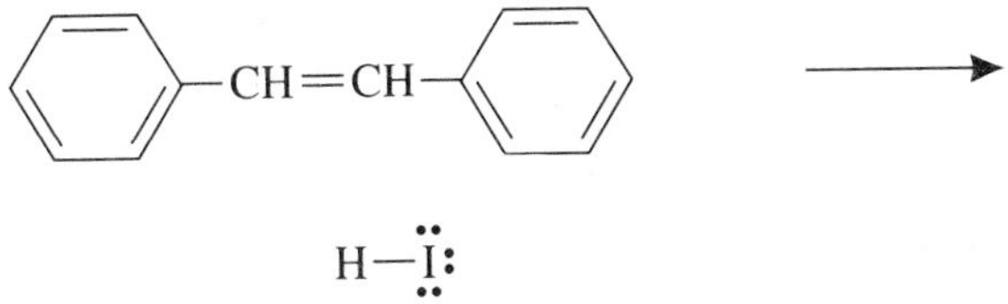

$C_6H_5-CH=CH-C_6H_5 \quad H-I: \longrightarrow$

supply arrows

* * * * *

When the carbon–carbon double bond is unsymmetrical, this type of reaction can lead to two different carbocation intermediates. The choice of which one is preferred is determined by the relative stabilities of the carbocations.

108. The protonation of propylene by hydrogen chloride can lead to two different carbocations. If as the pi bond breaks, the new sigma bond is made between C1 and the hydrogen atom, a **secondary carbocation** is formed.

$$\overset{3}{CH_3}-\overset{2}{CH}=\overset{1}{CH_2} \quad H-\ddot{Cl}: \longrightarrow CH_3-\overset{+}{CH}-CH_2-H \quad :\ddot{Cl}:^-$$

However, if the new sigma bond is made between C2 and the hydrogen atom, a **primary carbocation** is formed.

$$CH_3-CH=CH_2 \quad H-\ddot{Cl}: \longrightarrow$$

________________________

Because the transition state leading to the secondary carbocation is more stable, the first of these two products is formed preferentially.

109. When hydrogen iodide is allowed to react with isobutylene, one of two different carbocations can form. If the new sigma bond is formed between C1 and the hydrogen atom, a **tertiary carbocation** results.

$$\overset{3}{CH_3}-\overset{2}{\underset{}{C}}(CH_3)=\overset{1}{CH_2} \quad H-\ddot{I}: \longrightarrow$$

____________________ ________________________

supply arrows

However, if the new sigma bond is formed between C2 and the hydrogen atom, a primary carbocation is produced.

$$\underset{3}{CH_3}-\overset{CH_3}{\underset{2}{C}}=\underset{1}{CH_2} \qquad H-\ddot{\underset{\cdot\cdot}{I}}: \longrightarrow$$

supply arrows

The first of these two is preferred.

110. The reaction of bromine with styrene can yield a secondary, benzylic carbocation.

$$C_6H_5-CH=CH_2 \qquad :\ddot{\underset{\cdot\cdot}{Br}}-\ddot{\underset{\cdot\cdot}{Br}}: \longrightarrow$$

supply arrows

or, a primary carbocation,

$$C_6H_5-CH=CH_2 \qquad :\ddot{\underset{\cdot\cdot}{Br}}-\ddot{\underset{\cdot\cdot}{Br}}: \longrightarrow$$

supply arrows

The first reaction is preferred.

## Exercises

Here are some steps in organic mechanisms. In cases where arrows are supplied, write the products taking particular care about formal charge. In cases where products are given supply the appropriate arrows.

111. $CH_3-C(=\overset{+}{O}H)-\ddot{O}-C(CH_3)_3$ ⟶ ____________

112. $H_2C(-H)-\overset{+}{C}(CH_3)-CH_3$ + $H_2\ddot{O}$ ⟶ $H_2C=C(CH_3)-CH_3$ + $H_3\overset{+}{O}$

____________

113. $H-\ddot{O}-H$ + $Ph-C(-\ddot{O}H)=\ddot{N}-Ph$ + $H_3\overset{+}{O}$ ⟶ ____________

114. $Ph-CH(-\ddot{O}H)(-\ddot{O}:^-)$ + $Ph-CH=\ddot{O}$ ⟶ $Ph-C(=\ddot{O})-\ddot{O}H$ + $Ph-CH_2-\ddot{O}:^-$

115. (arenium ion: cyclohexadienyl cation bearing H and $NO_2$) + $:\ddot{O}^--SO_3H$ ⟶ ____________

116.

$Ph-CH_2-C(OH)(OH)-NH_2$ + $^-OH$ + $H_2O$ ⟶ $Ph-CH_2-C(=O)-OH$ + $H_2O$ + $NH_3$ + $^-OH$

---

117.

$CH_3-C(OH)(\overset{+}{O}H_2)-O-CH_2-CH_3$ + $H_3O^+$ + $H_2O$ ⟶ ________

## Rearrangements

The group of reactions in which rearrangements occur is one feature that lends organic chemistry its pizzazz. For some, discerning what happened during a rearrangement presents a stimulating challenge; for others, it is a complete mystery. In all cases the rearrangement can be figured out by pushing electrons properly and by carefully keeping track of the relative positions of the atoms.

The first, and most common, rearrangements encountered in the study of organic chemistry are 1,2-shifts in carbocations. In these reactions, a group (Y) one carbon atom away from a carbocation center moves, with its pair of electrons, to that center. The carbon atom from which that group migrated is left with a formal positive charge.

$$-C(Y)-\overset{+}{C}- \longrightarrow -\overset{+}{C}-C(Y)-$$

In this general example the pushable electrons are the sigma electrons of the C–Y bond, and the receptor is the positively charged carbon atom. The driving force for these 1,2-shifts is usually the formation of a more stable carbocation. The groups which most often take the role of Y are hydrogen, methyl, and aryl.

The 1-butyl cation suffers a 1,2-hydride shift to become the more stable 2-butyl cation. A neighboring hydrogen atom, with its pair of sigma electrons, migrates to the positively charged carbon atom, which becomes neutral. The neutral carbon atom from which the hydride ion migrated obtains a positive charge.

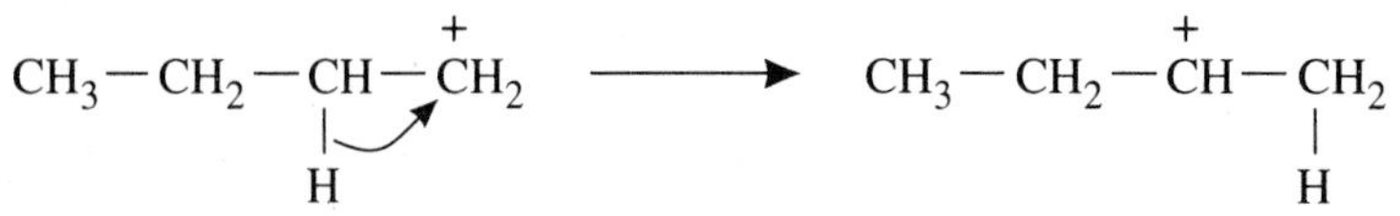

$$CH_3-CH_2-\underset{\displaystyle H}{\underset{|}{CH}}-\overset{+}{CH_2} \longrightarrow CH_3-CH_2-\overset{+}{CH}-\underset{\displaystyle H}{\underset{|}{CH_2}}$$

Notice, as you learned in Chapter 2, that *electrons* are pushed, not the positive charge.

118. The isobutyl cation rearranges via a hydride shift to the *t*-butyl cation.

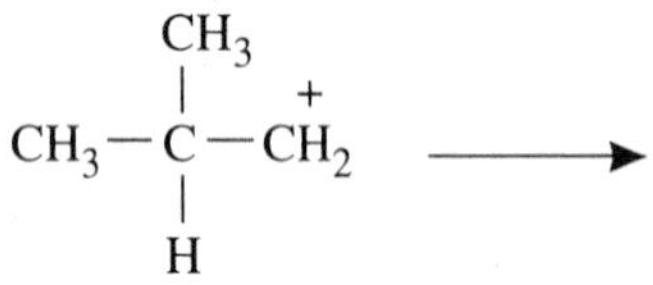

$$CH_3-\underset{\displaystyle H}{\underset{|}{\overset{\displaystyle CH_3}{\overset{|}{C}}}}-\overset{+}{CH_2} \longrightarrow$$

______________ ______________________

supply arrow

119. A 1,2-shift of a methyl group is illustrated in the rearrangement of the 3,3-dimethyl-2-butyl cation to the 2,3-dimethyl-2-butyl cation.

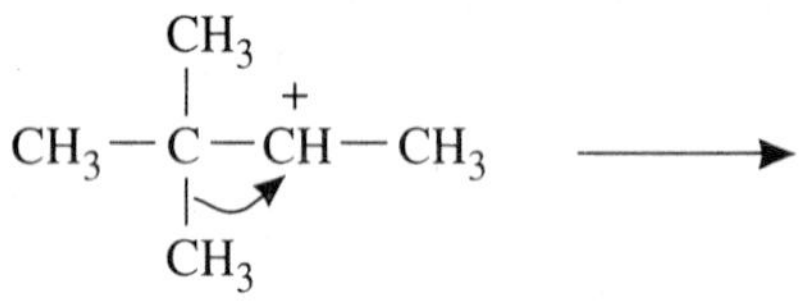

$$CH_3-\underset{\displaystyle CH_3}{\underset{|}{\overset{\displaystyle CH_3}{\overset{|}{C}}}}-\overset{+}{CH}-CH_3 \longrightarrow$$

______________________

120. Similarly,

$$CH_3-\underset{\displaystyle CH_3}{\underset{|}{\overset{\displaystyle CH_3}{\overset{|}{C}}}}-\overset{+}{CH_2} \longrightarrow$$

______________ ______________________

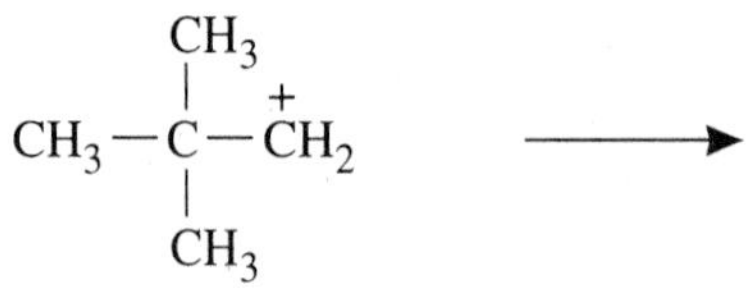

supply arrow

* * * * *

In each of the cases above the rearrangement resulted in the formation of a more stable cation, that is, a secondary and tertiary cation are formed from a primary, and another tertiary cation is formed from a secondary. Very often one must choose the most likely rearrangement from among several alternatives.

121. The 1-propyl cation can rearrange via either a 1,2-hydride shift or a 1,2-methyl shift. The 1,2-hydride shift leads to the more stable isopropyl cation.

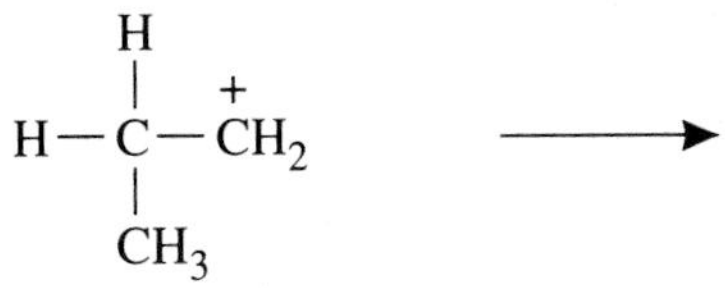

$$\mathrm{H{-}\underset{\underset{CH_3}{|}}{\overset{\overset{H}{|}}{C}}{-}\overset{+}{C}H_2} \longrightarrow$$

______________ ______________

supply arrow

A 1,2-methyl shift is not as likely because it would yield a primary cation as the product of rearrangement.

122. Show the most likely rearrangement for the 2-methyl-1-butyl cation.

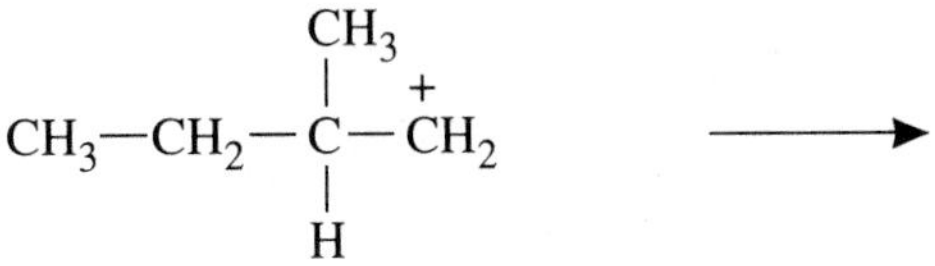

$$\mathrm{CH_3{-}CH_2{-}\underset{\underset{H}{|}}{\overset{\overset{CH_3}{|}}{C}}{-}\overset{+}{C}H_2} \longrightarrow$$

______________ ______________

supply arrow

123. In the cation below there are three different groups (hydrogen on $C_1$, hydrogen on $C_3$, or methyl on $C_3$) that can shift to give three possible products of rearrangement. Show the most likely rearrangement.

$$\underset{4}{\mathrm{CH_3}}{-}\underset{\underset{H}{|3}}{\overset{\overset{CH_3}{|}}{\mathrm{C}}}{-}\underset{2}{\overset{+}{\mathrm{C}}\mathrm{H}}{-}\underset{\underset{H}{|1}}{\mathrm{CH_2}} \longrightarrow$$

______________ ______________

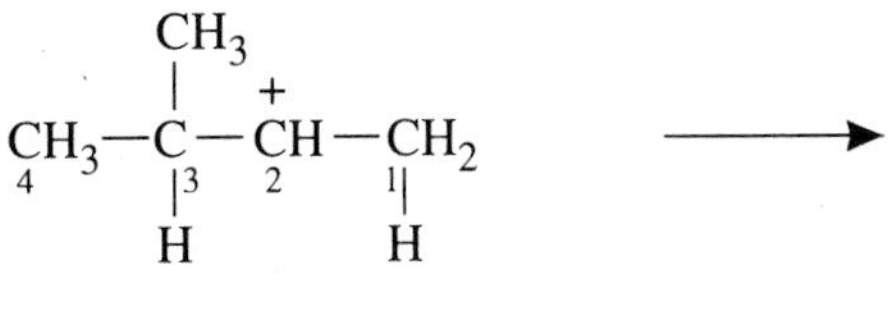

supply arrow

124. A 1,2-methyl shift occurs in the pinacol–pinacolone rearrangement.

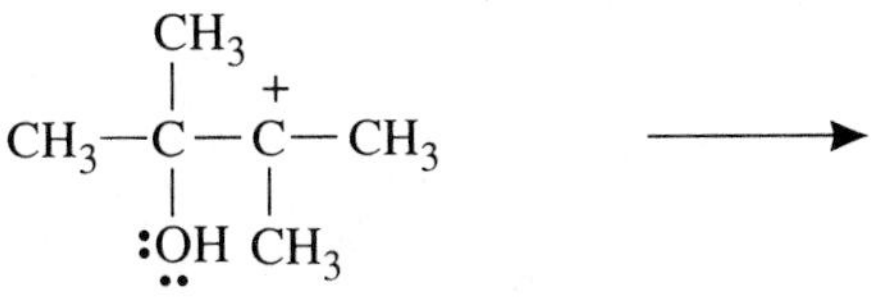

$$\mathrm{CH_3{-}\underset{\underset{:\ddot{O}H}{|}}{\overset{\overset{CH_3}{|}}{C}}{-}\underset{\underset{CH_3}{|}}{\overset{+}{C}}{-}CH_3} \longrightarrow$$

______________ ______________

supply arrow

125. Groups can migrate from one position on a ring to another.

$CH_3$

$CH_3$

+ H

supply arrow

126. When a phenyl group is found next to a cationic center, there is a strong possibility that it will migrate.

Ph—C—$\overset{+}{C}$H—Ph (with Ph below) → Ph—$\overset{+}{C}$—CH—Ph (with Ph below)

supply arrow

127. Phenyl groups can migrate from one position on a ring to another.

Ph

+ H

supply arrow

128. Phenyl groups migrate in the pinacol–pinacolone type rearrangements.

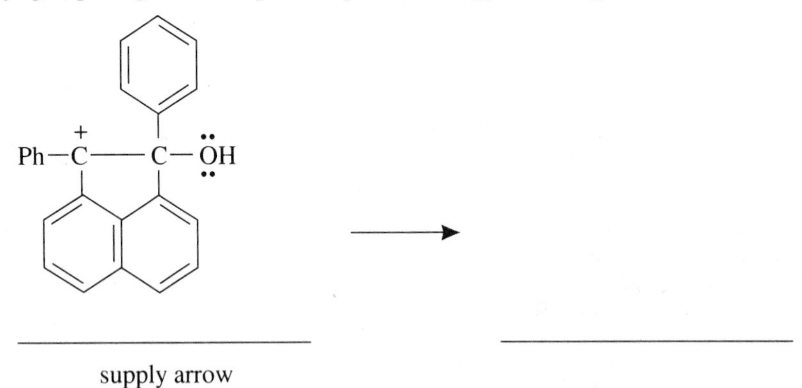


supply arrow

129. Electron releasing substituents on the aryl ring tend to make that ring migrate in preference to an unsubstituted phenyl group. In the reaction below, the *para*-methoxyphenyl group migrates in preference to phenyl.

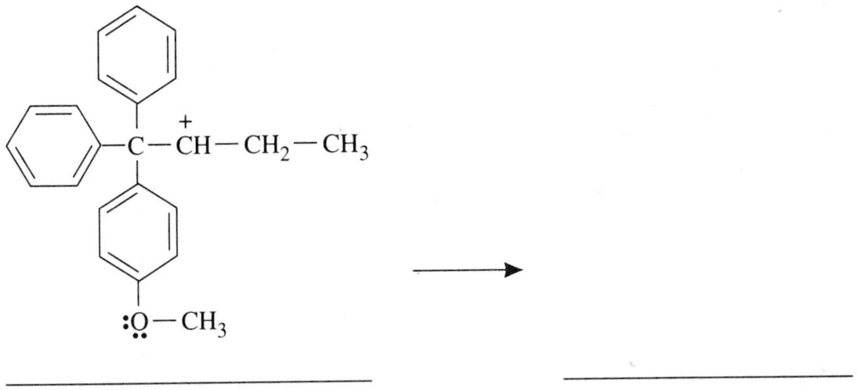

$(C_6H_5)_2(p\text{-}CH_3O\text{-}C_6H_4)C-\overset{+}{C}H-CH_2-CH_3 \longrightarrow$

$:\ddot{O}-CH_3$

______________________ ______________________

supply arrow

130. Electron attracting groups have the opposite effect so that phenyl migrates in preference to *para*-chlorophenyl.

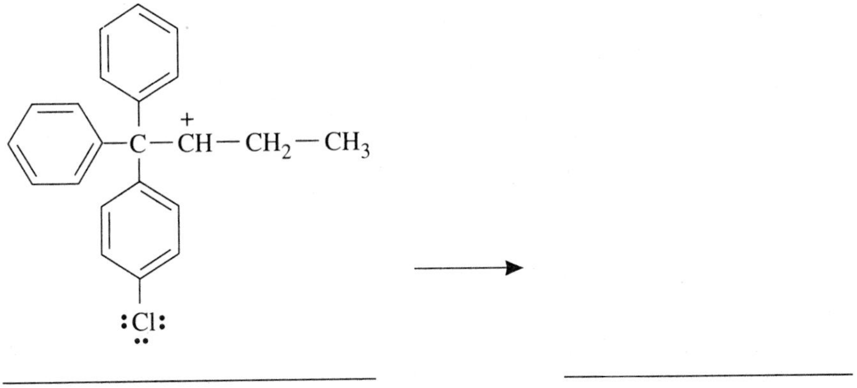

$(C_6H_5)_2(p\text{-}Cl\text{-}C_6H_4)C-\overset{+}{C}H-CH_2-CH_3 \longrightarrow$

$:\ddot{Cl}:$

______________________ ______________________

supply arrow

## Exercises

Here are some carbocations that can be expected to rearrange via 1,2-shifts. Show the rearrangement which is most probable.

131. $CH_3-C(CH_3)_2-\overset{+}{C}H-C(H)(CH_3)-H \longrightarrow$

______________________ ______________________

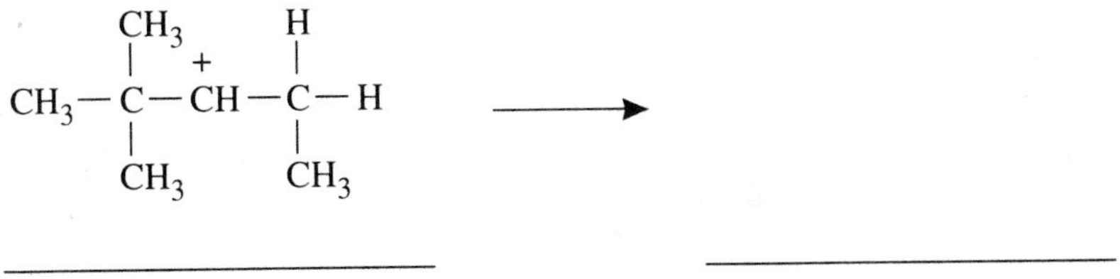

132.

$(C_6H_5)_3C-\overset{+}{C}H-C(CH_3)_3$ ⟶

133.

134.

135:

$Ph-C(C_6H_4CH_3)(C_6H_5)-\overset{+}{C}H-Ph$ ⟶

136.

* * * * *

Another 1,2-shift involving carbocations is very similar to the rearrangements we have already seen. However, it often tortures students because it involves ring expansion or ring contraction that is difficult to see at first. In the general example below, a carbocation outside the ring is a receptor for a migrating methylene group. The result of such a migration is, inevitably, a ring containing one more carbon than the original ring. The positive charge is now on a carbon atom in the ring.

The cyclohexylmethyl cation rearranges by expanding to the cycloheptyl cation. The author has found it useful to show the rearrangement of the bonds while keeping the positions of the atoms constant. Then, after the rearrangement has been shown, the structure can be altered to a more familiar form.

137. The cyclopropyl methyl cation suffers a similar rearrangement.

138. Similarly, the cyclobutylmethyl cation undergoes a ring expansion.

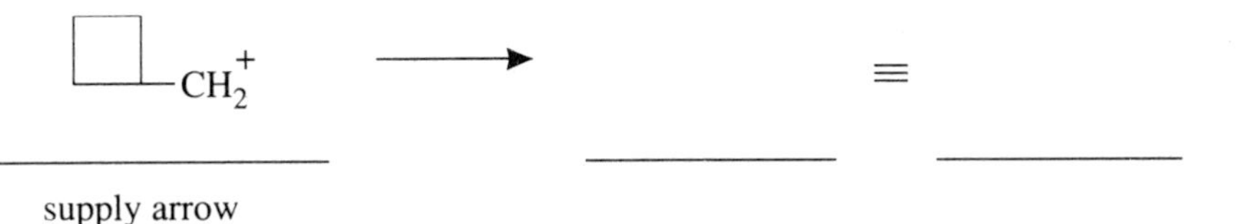

supply arrow

139. Another example.

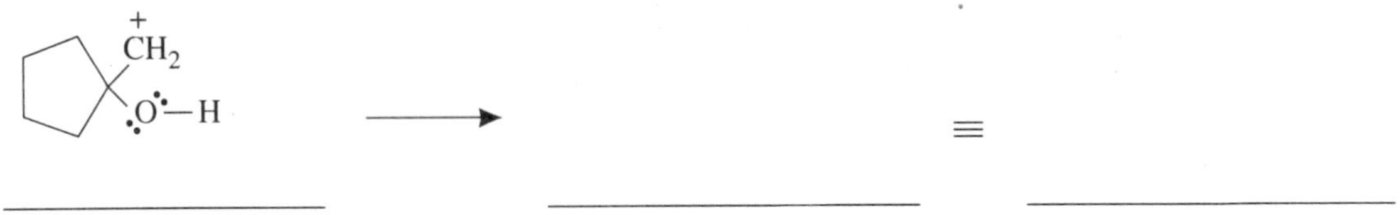

$\overset{+}{CH_2}$ ... $\ddot{O}$—H → ≡

supply arrow

* * * * *

Alternatively, the carbocation center can be inside the ring. In this case the migration of a methylene group leads to a ring contraction. The ring will contain one fewer member than the original ring, and the positive charge will be on a carbon atom outside the ring.

The 2,2-dimethylcyclohexyl cation undergoes a ring-contracting rearrangement. Once again it is useful to keep the atoms in the same position as the bonds are rearranged.

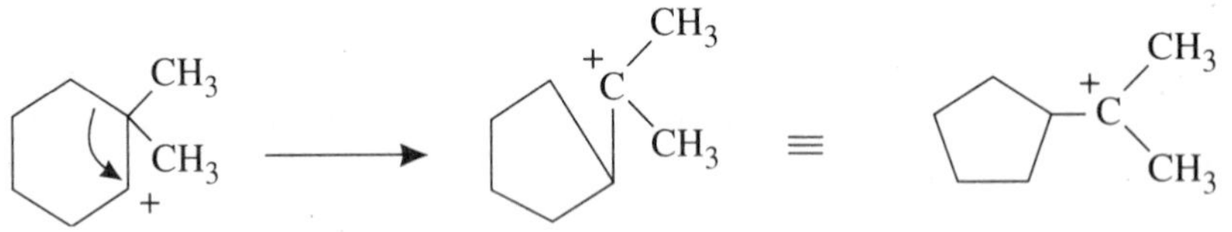

140. The cyclopentyl cation will contract to the cyclobutylmethyl cation.

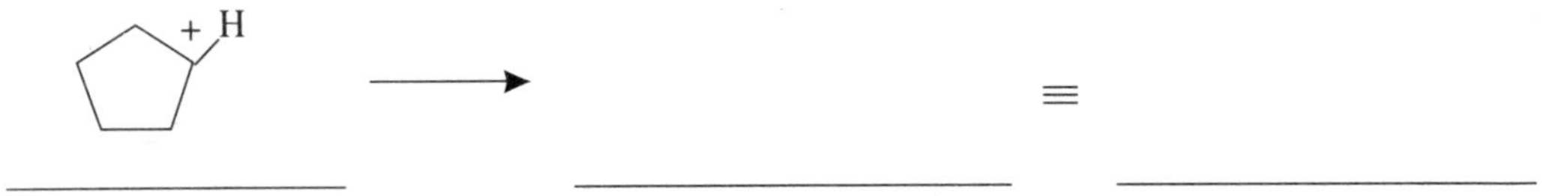

supply arrow

141. Another example.

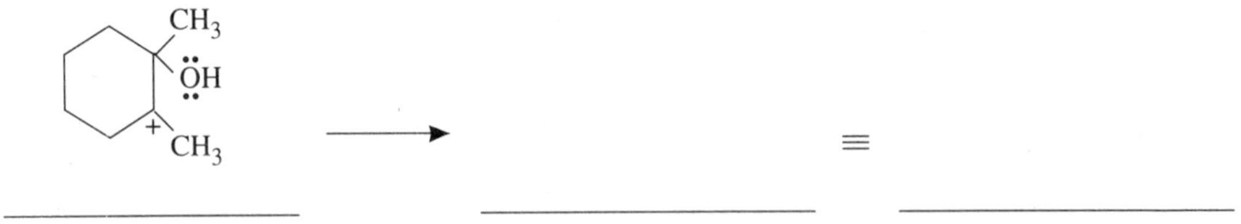

supply arrow

142. A slight variation on this reaction occurs when cyclopropyl cations rearrange to form allyl cations. Scrutiny of this process will show it to be like a ring contraction, the double bond being comparable to a two-membered ring.

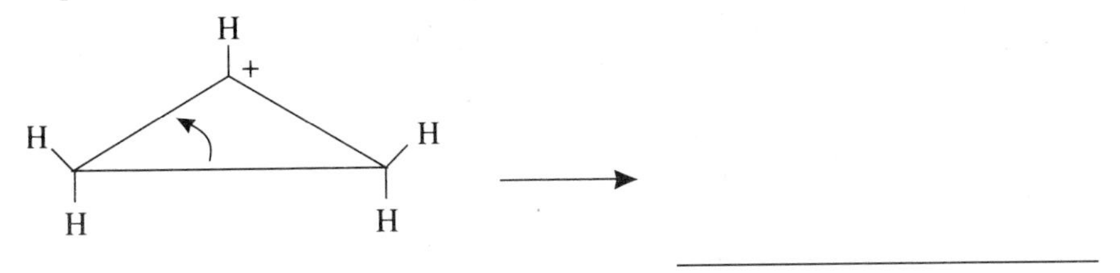

143. Another example.

## Ring Closures

The trick of holding atoms in place while pushing electrons works well in another class of reactions that sometimes vexes students. Consider this remarkable transformation.

$$\underset{5}{H_2C}{=}\underset{4}{CH}{-}\underset{3}{CH_2}{-}\underset{2}{CH_2}{-}\underset{1}{CH_2}{-}\ddot{\underset{..}{O}}{-}H \xrightarrow[H_2O]{Br_2} \text{(5-membered ring containing } \ddot{O}\text{, bearing } CH_2{-}\ddot{\underset{..}{Br}}\text{:)}$$

After rejecting the notion that a miracle has occurred, the wise student works on three premises: familiar chemistry is at work, most of the atoms in the starting material will appear in the product, and a minimum number of bonds have been broken and made.

Study how the product relates to the starting materials. There are five carbons and one oxygen in each. Connecting the oxygen with C4 will give a 5-membered ring. Carbon-4 is part of a double bond. Bromine reacts with double bonds.

$$H_2C{=}CH{-}CH_2{-}CH_2{-}CH_2{-}\ddot{\underset{..}{O}}{-}H \quad :\ddot{\underset{..}{Br}}{-}\ddot{\underset{..}{Br}}: \longrightarrow \underset{|\,:\ddot{Br}:}{CH_2}{-}\overset{+}{C}H{-}CH_2{-}CH_2{-}CH_2{-}\ddot{\underset{..}{O}}{-}H$$

Now there is a way for oxygen to bind to C4. Note that, temporarily, all the atoms are held in place.

$CH_2-\overset{+}{C}H-CH_2-CH_2-CH_2-\ddot{O}-H \longrightarrow CH_2-CH-CH_2-CH_2-CH_2-\overset{+}{O}-H$ (with :Br: on the first $CH_2$)

Now the structure with the bizarre, elongated bond is converted to the more realistic and familiar form.

$CH_2-CH-CH_2-CH_2-CH_2-\overset{+}{O}-H \equiv$ (ring: $:\overset{+}{O}-H$, $CH_2$, $CH-CH_2-\ddot{B}r:$, $CH_2-CH_2$)

Loss of a proton from oxygen gives the product.

Here are some ring-closure reactions. Some are homolytic; others are heterolytic. Where arrows are supplied, draw the products. Where products are given, supply arrows.

144. $\ddot{O}:$ $CH_3$ $:\ddot{O}$ $-CH_2-CH_2-\overset{\|}{C}-\overset{-}{\ddot{C}H_2}$ $\longrightarrow$ $\ddot{O}:$

______________________

145. $CH_2{=}CH-CH_2-CH_2-CH_2-\dot{C}H_2 \longrightarrow$ $\dot{C}H_2$

______________________

146. $\overset{+}{\ddot{O}}-H$ $H-\overset{\|}{C}-CH_2-CH_2-CH_2-CH_2-CH{=}C(CH_3)_2 \longrightarrow$

______________________

147. $:\overset{-}{\ddot{O}}-CH_2-CH_2-CH_2-CH_2-\ddot{B}r: \longrightarrow$ $\ddot{O}$ $+ :\ddot{B}r:^-$

______________________

148. $:\ddot{\underset{..}{Cl}}-CH_2-CH_2-CH_2-CH_2-\overset{-}{\underset{..}{C}}(COOR)_2$ ⟶

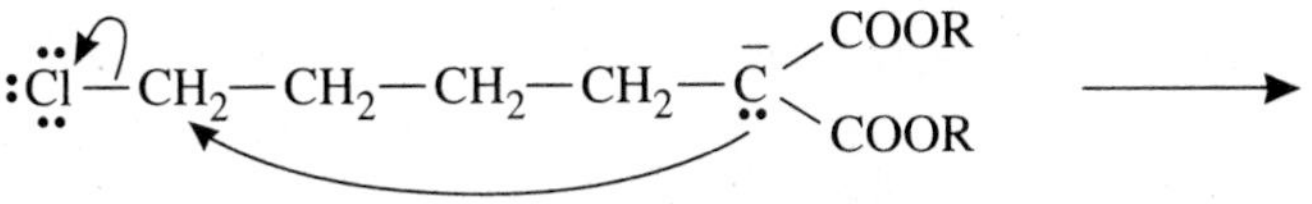

149. $CH_2=CH-\underset{\underset{:\ddot{O}}{\|}}{C}-\ddot{\underset{..}{O}}-(CH_2)_9-\dot{C}H_2$ ⟶

______________________________

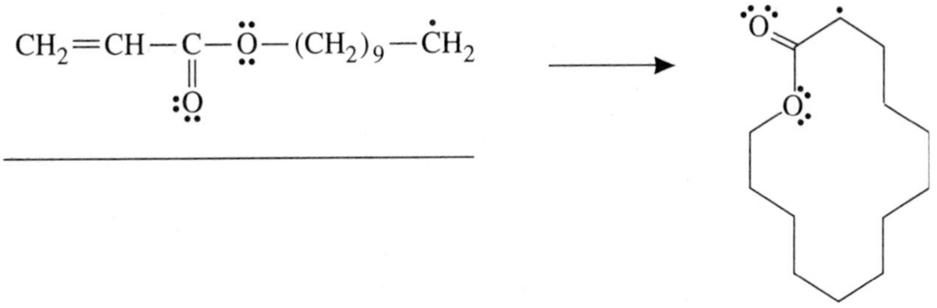

150. $(CH_3)_2C=CH-CH_2-CH_2-CH_2-\overset{+}{C}(CH_3)_2$ ⟶

______________________________

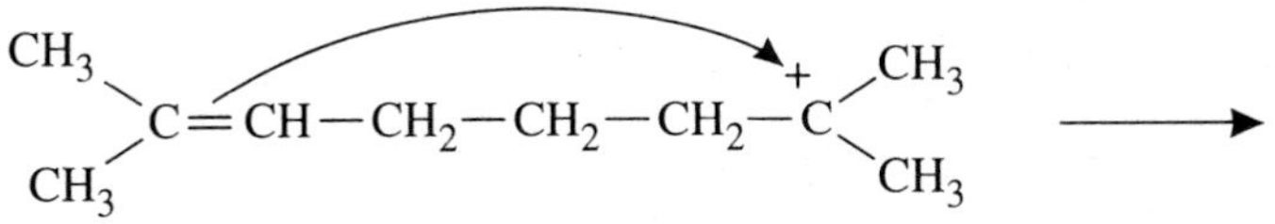

151. $H-\overset{+}{O}=CH-CHOH-CHOH-CHOH-\underset{H}{C}H-\ddot{\underset{..}{O}}-H$ (with terminal $CH_2OH$) ⟶ $CH_2OH$, $CH-\overset{+}{O}:$ (H), $HOCH$, $CHOH$, $CH-CH$, $OH$ $OH$

______________________________

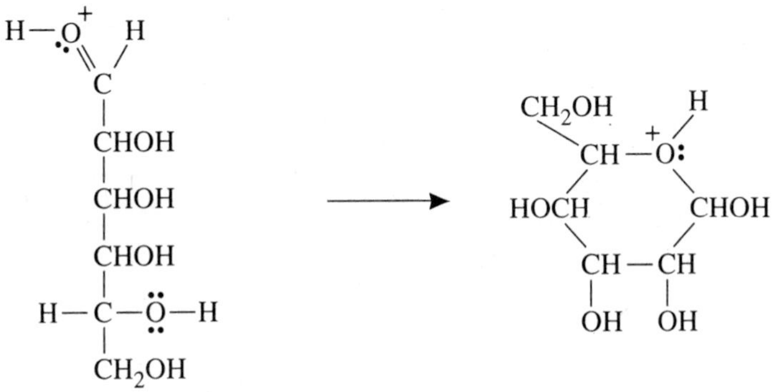

152. $\cdot CH_2-(CH_2)_4-CH=CH_2$ ⟶

______________________________

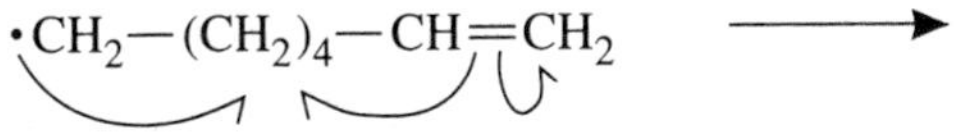

153. $R-\ddot{\underset{..}{O}}-\underset{\underset{:\ddot{O}:}{\|}}{C}-\overset{-}{\underset{..}{C}}H-CH_2-CH_2-CH_2-\underset{\underset{:\ddot{O}:}{\|}}{C}-\ddot{\underset{..}{O}}-R$ ⟶ $R-\ddot{O}$, $\ddot{O}:^-$, $COOR$

______________________________

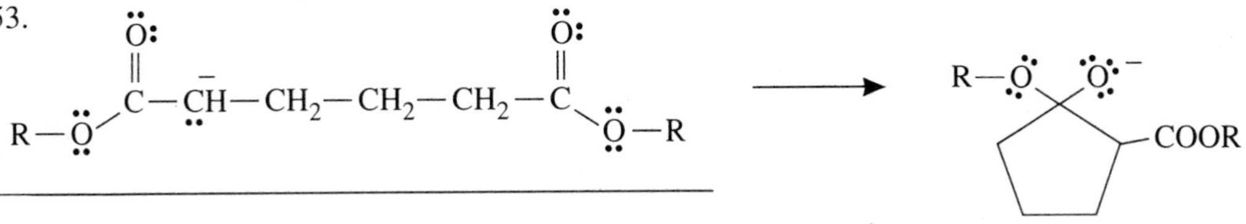

# Answers

## Chapter 3

1. chloride, carbon–chlorine, chlorine, negative, positive, isopropyl,

$$CH_3-\overset{CH_3}{\underset{H}{C^+}} \qquad :\ddot{Cl}:^-$$

2. bromide, carbon–bromine, bromine, negative, positive, bromide, cyclohexyl,

$$C_6H_{11}-\ddot{Br}: \longrightarrow C_6H_{11}^+ \quad :\ddot{Br}:^-$$

3.

$$CH_3-\overset{CH_3}{\underset{CH_3}{C}}-\overset{+}{\ddot{O}}(H)-H \longrightarrow CH_3-\overset{CH_3}{\underset{CH_3}{C^+}} \quad :\ddot{O}(H)-H$$

4. oxygen, neutral, positive,

$$CH_3-\ddot{O}-CH_2-\overset{+}{\ddot{O}}(H)-\overset{:\ddot{O}:}{\overset{\|}{C}}-CH_3 \longrightarrow CH_3-\ddot{O}-CH_2^+ \quad :\ddot{O}(H)-\overset{:\ddot{O}:}{\overset{\|}{C}}-CH_3$$

5. oxygen positive, neutral, positive,

$$CH_3-\overset{CH_3}{C}-\overset{+}{\ddot{O}}-H \text{ (bridged with) } CH_3-\underset{CH_3}{C}$$

6.

$$C_6H_5-CH_2-\ddot{Br}: \longrightarrow C_6H_5-CH_2^+ \quad :\ddot{Br}:^-$$

7. $Cl{-}CH(C_6H_5)_2 \longrightarrow Cl^- \quad {}^+CH(C_6H_5)_2$

8. $C_6H_{11}{-}\overset{+}{O}(H){-}SO_2{-}C_6H_4{-}CH_3 \longrightarrow C_6H_{11}^+ \quad HO{-}SO_2{-}C_6H_4{-}CH_3$

9. $CH_3{-}O{-}C_6H_4{-}CH_2{-}Cl \longrightarrow CH_3\,O{-}C_6H_4{-}\overset{+}{C}H_2 \quad Cl^-$

10. $Ph_2C(\overset{+}{O}H){-}CPh_2 \longrightarrow Ph_2C(OH){-}C^+Ph_2$

11. $C_6H_5{-}C(=O){-}\overset{+}{O}(H){-}CH_2{-}O{-}CH_2{-}CH_3 \longrightarrow C_6H_5{-}C(=O){-}O{-}H \quad {}^+CH_2{-}O{-}CH_2CH_3$

12. $H{-}\overset{+}{O}(CH(CH_3)){-}CH(CH_3) \longrightarrow H{-}O{-}CH(CH_3){-}\overset{+}{C}H{-}CH_3$

13.

$$CH_3-C(CH_3)_2-\overset{+}{O}(H)-C(=O)-C_6H_5 \longrightarrow CH_3-\overset{+}{C}(CH_3)_2 \quad :\ddot{O}(H)-C(=O)-C_6H_5$$

14.

$$\text{(cyclopentyl)}-\overset{+}{O}(H)-SO_2-C_6H_4-Br: \longrightarrow H-\ddot{O}:-SO_2-C_6H_4-Br: \quad + \text{cyclopentyl cation}$$

15. unshared, bromide, carbon, bromine, negative, neutral

$$CH_3-CH_2-\overset{+}{C}H(CH_3) \;\; :\ddot{Br}:^- \longrightarrow CH_3-CH_2-CH(CH_3)-\ddot{Br}:$$

16.

$$Ph-\overset{+}{C}H(CH_3) \;\; :\ddot{O}(H)-CH_2-CH_3$$

17. unshared, nitrogen, carbon, sigma, positive, positive, neutral,

$$\text{cyclohexyl}^+ \;\; :N(H)(H)-H \longrightarrow \text{cyclohexyl}-\overset{+}{N}(H)(H)-H$$

18.

$$C_6H_5-\overset{+}{C}H_2 \;\; :\ddot{F}:^- \longrightarrow C_6H_5-CH_2-\ddot{F}:$$

19.

$$\text{cyclohexyl}^+(H) \;\; {}^-:\ddot{O}-C(=\ddot{O}:)-CH_3 \longrightarrow \text{cyclohexyl}(H)-\ddot{O}-C(=\ddot{O}:)-CH_3$$

20.

$$CH_3-CH_2-\overset{+}{C}H(CH_3) \;\; :\ddot{O}(H)-CH_3 \longrightarrow CH_3-CH_2-CH(CH_3)-\overset{+}{\ddot{O}}(H)-CH_3$$

21.

$$Ph-\overset{+}{C}H(Ph) \leftarrow :N(H)_2-CH_3 \longrightarrow Ph-CH(Ph)-\overset{+}{N}(H)_2-CH_3$$

22.

$$CH_2-\overset{+}{C}H_2 \text{ with } :\ddot{O}(H)- \text{ on the first } CH_2 \longrightarrow \text{ring of } CH_2-CH_2 \text{ and } :\overset{+}{O}-H$$

23. hydroxide, chloride, unshared, sigma, chlorine, carbon, chlorine, neutral, negative,

$$H-\ddot{\underset{..}{O}}:^- \curvearrowright CH_2(CH_3)-\ddot{\underset{..}{Cl}}: \longrightarrow H-\ddot{\underset{..}{O}}-CH_2(CH_3) \quad :\ddot{\underset{..}{Cl}}:^-$$

24.

$$H-\ddot{\underset{..}{O}}:^- \curvearrowright CH_2(C_6H_5)-\ddot{\underset{..}{Br}}: \longrightarrow H-\ddot{\underset{..}{O}}-CH_2(C_6H_5) \quad :\ddot{\underset{..}{Br}}:^-$$

25. hydrogen, oxygen,

$$C_6H_5-\ddot{\underset{..}{O}}:^- \qquad H-\ddot{\underset{..}{O}}-H$$

26. unshared, nitrogen, sigma, carbon, hydrogen, carbon, neutral, negative,

$$H-C\equiv C-H \leftarrow :\ddot{N}^-(H)-H \longrightarrow H-C\equiv C:^- \quad H-\ddot{N}(H)-H$$

27.

$$CH_3-CH_2-\ddot{\underset{..}{O}}:^- \rightarrow H-CH_2-C(=\ddot{O}:)H \longrightarrow CH_3-CH_2-\ddot{\underset{..}{O}}-H \quad ^-:CH_2-C(=\ddot{O}:)H$$

28. *n*-propyl, unshared, carbon, sigma, oxygen, oxygen, carbon, oxygen, neutral, negative,

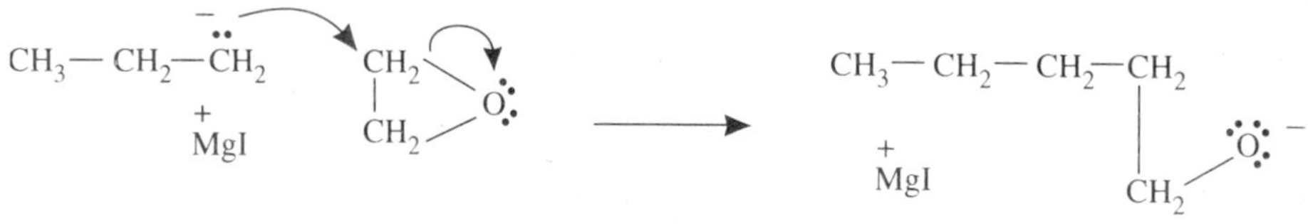

29. ethoxide, unshared, oxygen, carbon, sigma, oxygen, oxygen, carbon, oxygen, neutral, negative,

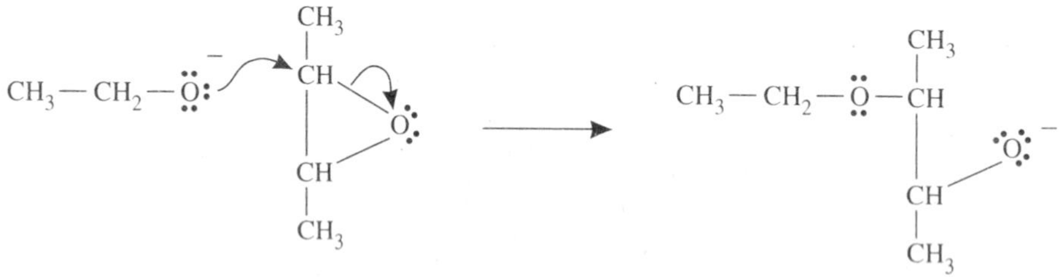

30. cyclohexanol, oxygen, hydronium, unshared, hydrogen, sigma, oxygen, oxygen, hydrogen, oxygen, positive, neutral,

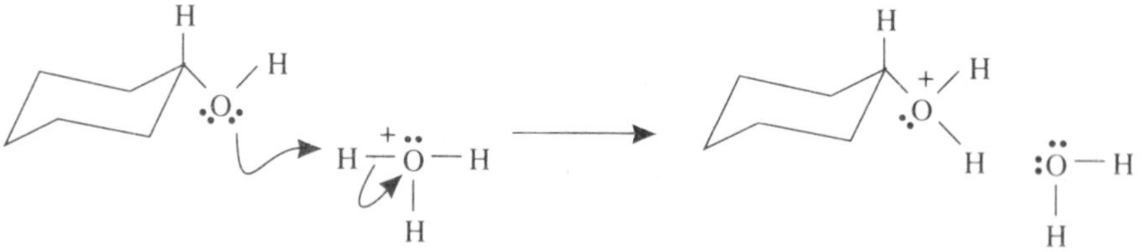

31. oxygen, anilinium, unshared, hydrogen, sigma, nitrogen, nitrogen, hydrogen, nitrogen, neutral, positive,

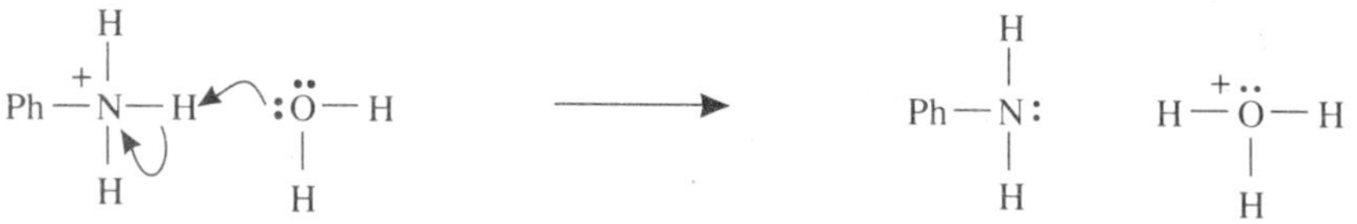

32.

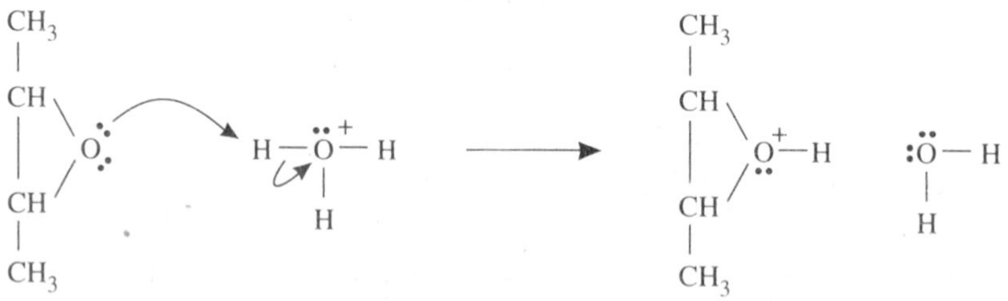

33. $H_2\ddot{O}:$ + $CH_3-CH(-CH-CH_3)-\overset{+}{O}H$ (ring-bridged) ⟶ $H_2\overset{+}{O}-CH(CH_3)-CH(CH_3)-\ddot{O}-H$

$$H_2\ddot{O}: \;+\; CH_3-\underset{\;}{CH}\!-\!CH-CH_3 \;(\overset{+}{O}-H) \longrightarrow H_2\overset{+}{O}-CH(CH_3)-CH(CH_3)-\ddot{O}-H$$

34.

$$H_2\ddot{O}: \;+\; H_2\overset{+}{O}-CH(CH_3)-CH(CH_3)-\ddot{O}-H \longrightarrow H-\overset{+}{\underset{H}{\ddot{O}}}-H \;\; H-\ddot{O}-CH(CH_3)-CH(CH_3)-\ddot{O}-H$$

35.

$$(CH_3)_2C=\overset{+}{\ddot{O}}-H \qquad :\ddot{O}(H)-H$$

36.

$$CH_3-CH_2-\overset{+}{\ddot{O}}(Ph)-H \;\; H-:\ddot{O}-H$$

37.

$$C_6H_5-\ddot{O}-CH_2-CH_3 \qquad :\ddot{I}:^-$$

38.

$$CH_3-C(=\ddot{O})-\ddot{O}-CH_2-CH_3 \qquad H-\ddot{O}-SO_3H$$

39.

$$H(Ph)C=\ddot{O}: \qquad H-\ddot{O}-C(=\ddot{O}:)-CH_3$$

40. $H_3C$, $\ddot{C}H^-$, $H_3C$, $\overset{+}{MgBr}$; $CH_3$–CH–CH–$CH_3$ (epoxide, $\ddot{O}$)

41. $CH_3-CH_2$, $CH_3-CH_2$, $+N$, H, H; $:\ddot{O}^--C(=\ddot{O}:)-CH_3$

42. $CH_3-C(CH_3)_2-\ddot{O}:^-$ → $CH_3-\ddot{Br}:$

43. $CH_3-CH_2\cdot$ $H-\ddot{Cl}:$

44. (cyclohexane)–H, H $\cdot\ddot{Cl}:$ ⟶ (cyclohexyl radical)–H

45. $H_3C-\ddot{Br}:$ $\cdot\ddot{Br}:$

46. $Cl_3Si-H$ $\cdot\ddot{O}-C(=\ddot{O}:)-Ph$ ⟶ $Cl_3Si\cdot$

47. $CH_3-C(CN)(CH_3)\cdot$ $H-Sn-R_3$ ⟶ $CH_3-C(CN)(CH_3)-H$ + $\cdot SnR_3$

48. $CH_3-C(H)(CH_3)-H$ $\cdot\ddot{Cl}:$ ⟶ $CH_3-C(H)(CH_3)\cdot$

49. 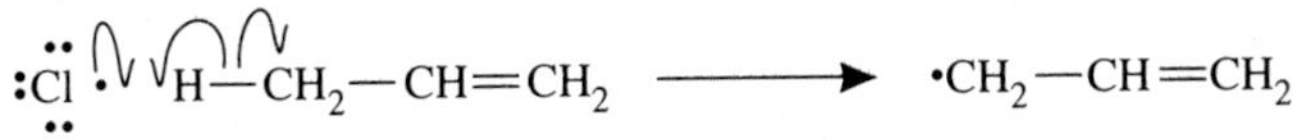 $:\ddot{Cl}\cdot\ H{-}CH_2{-}CH{=}CH_2 \longrightarrow \cdot CH_2{-}CH{=}CH_2$

50. 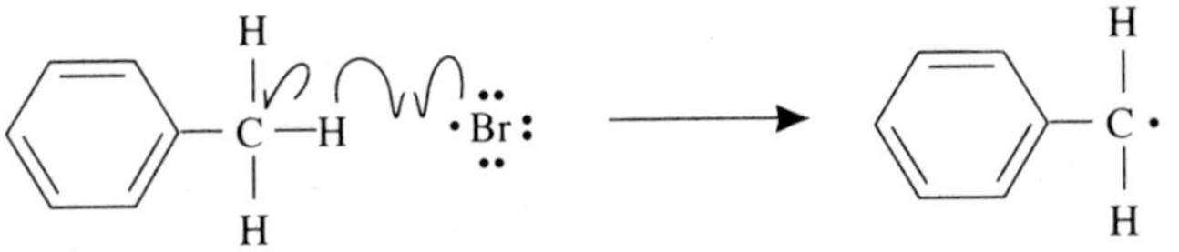 $C_6H_5{-}CH_2{-}H\ \ \cdot\ddot{Br}: \longrightarrow C_6H_5{-}\dot{C}H_2$

51.  $CH_2{=}CH{-}(CH_2)_3{-}CH_2{-}\ddot{Br}:\ \ \cdot SnR_3 \longrightarrow CH_2{=}CH{-}(CH_2)_3{-}\dot{C}H_2 + :\ddot{Br}{-}SnR_3$

52. 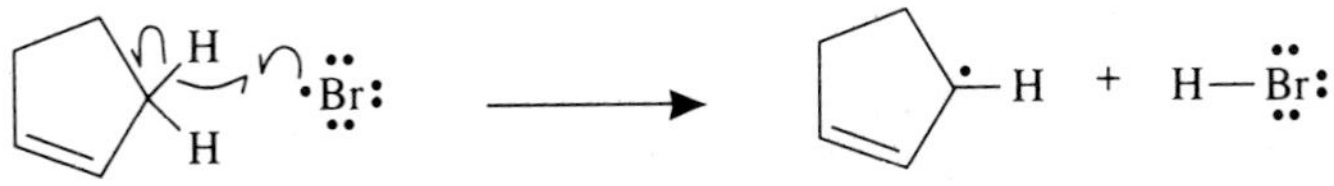 cyclopentene–H + $\cdot\ddot{Br}:$ $\longrightarrow$ cyclopentenyl radical (•–H) + $H{-}\ddot{Br}:$

53. 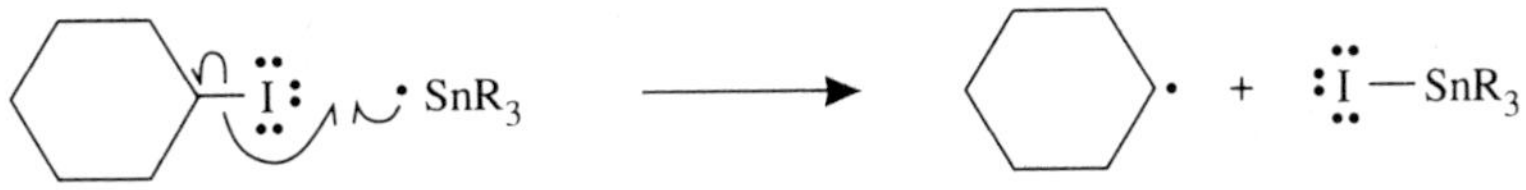 cyclohexyl–$\ddot{I}:$ + $\cdot SnR_3$ $\longrightarrow$ cyclohexyl• + $:\ddot{I}{-}SnR_3$

54. 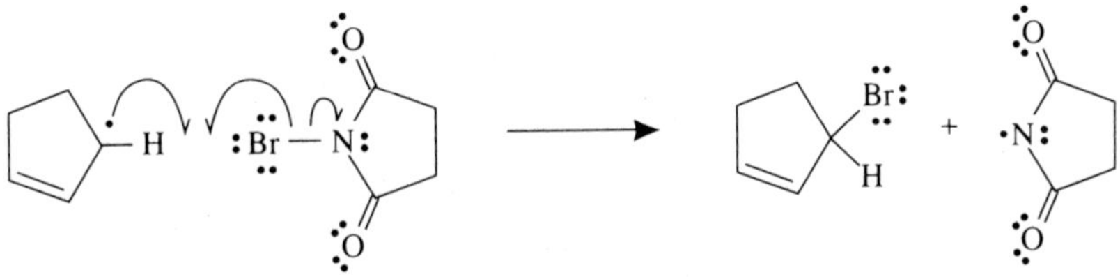 cyclopentenyl• (–H) + $:\ddot{Br}{-}N:$ (succinimide) $\longrightarrow$ bromocyclopentene ($\ddot{Br}:$, H) + •N: (succinimidyl radical)

55. hydroxide, unshared, oxygen, oxygen, negative

56. hydroxide, unshared, pi, oxygen, oxygen, carbon, oxygen, neutral, negative,

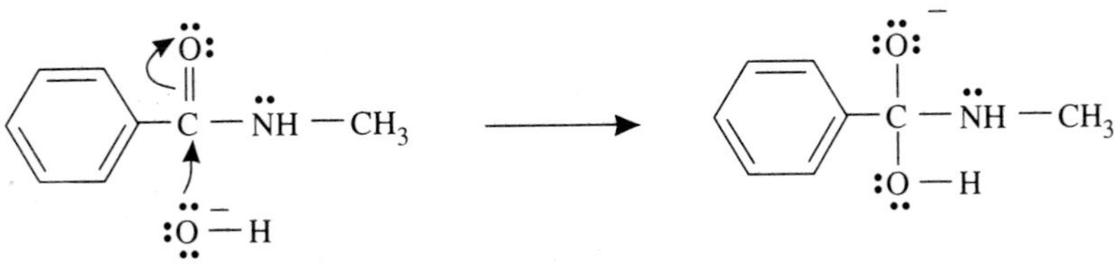

$C_6H_5{-}C(=\ddot{O}:){-}\ddot{N}H{-}CH_3 + :\ddot{O}^-{-}H \longrightarrow C_6H_5{-}C(:\ddot{O}:^-)(:\ddot{O}{-}H){-}\ddot{N}H{-}CH_3$

57. 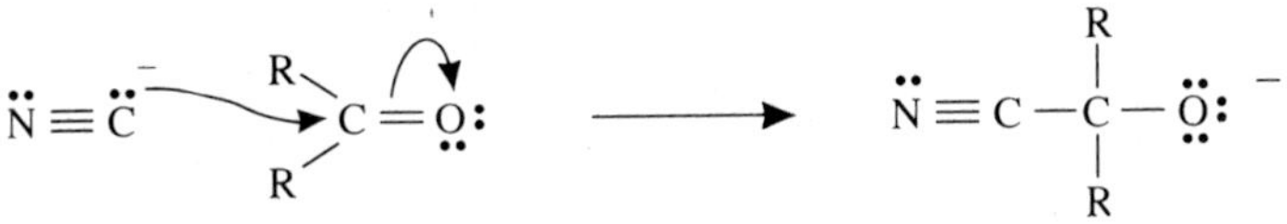 $:N{\equiv}\ddot{C}^- + R_2C{=}\ddot{O}: \longrightarrow :N{\equiv}C{-}CR_2{-}\ddot{O}:^-$

58.

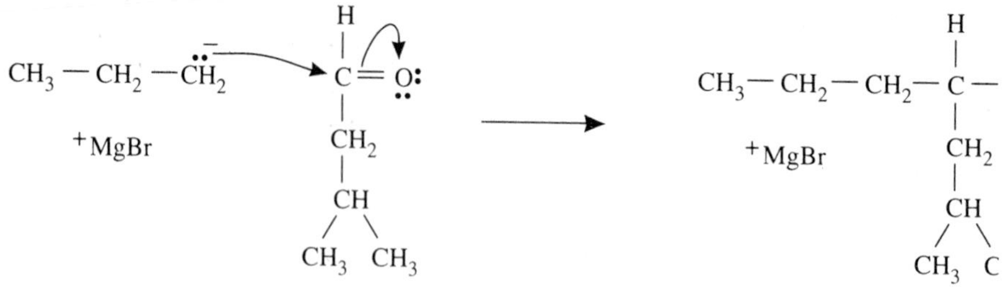

$CH_3-CH_2-\ddot{C}H_2^-$ $^+MgBr$ → $CH_3-CH_2-CH_2-C(H)(CH_2CH(CH_3)_2)-$ $^+MgBr$

59.

$C_6H_5-C\equiv N:$ + $:\ddot{O}^--H$ → $C_6H_5-C(OH)=\ddot{N}:^-$

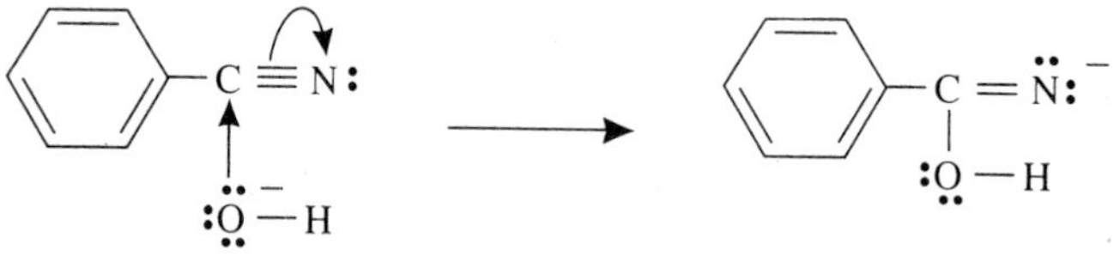

60. carbon, oxygen, positive, neutral

61. unshared, oxygen, pi, oxygen, carbon, oxygen, positive, neutral

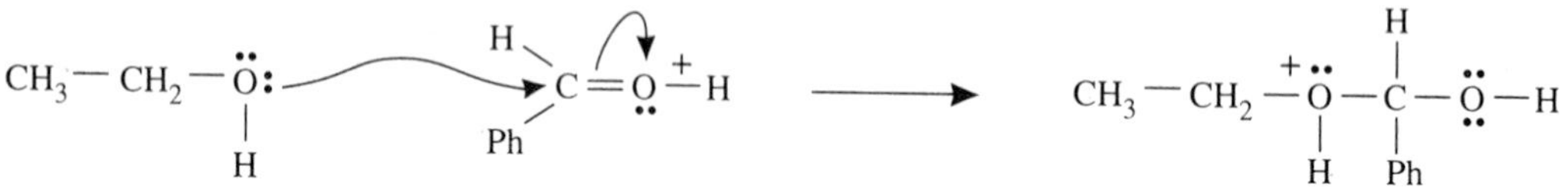

$CH_3-CH_2-\ddot{O}(H):$ + $H(Ph)C=\ddot{O}^+-H$ → $CH_3-CH_2-\overset{+}{O}(H)-C(H)(Ph)-\ddot{O}-H$

62.

$Ph-CH_2-C\equiv N^+-H$ + $:\ddot{O}(H)-H$ → $Ph-CH_2-C(\overset{+}{O}H_2)=\ddot{N}-H$

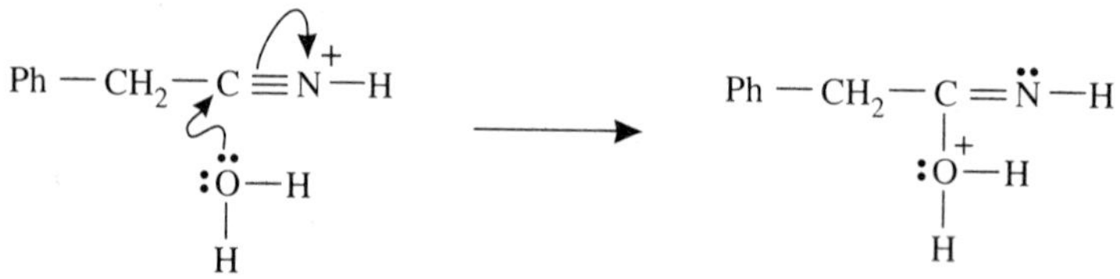

63. pi, carbon, carbon, neutral, positive,

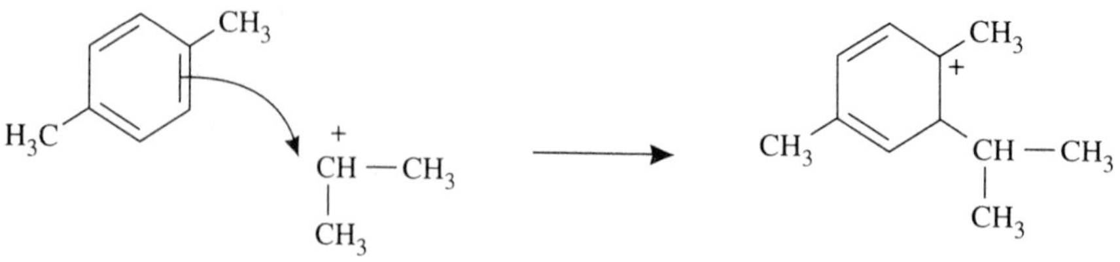

$H_3C-C_6H_4-CH_3$ + $\overset{+}{C}H(CH_3)-CH_3$ → $CH_3-C_6H_4^+(CH_3)-CH(CH_3)-CH_3$

64.

$CH_3-C(CH_3)=CH_2$ + $^+C(CH_3)_3$

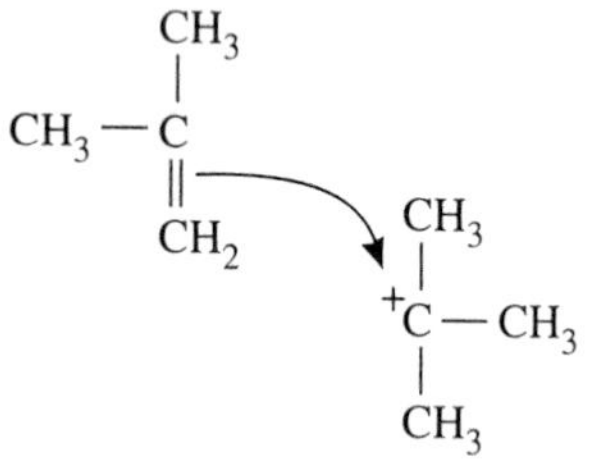

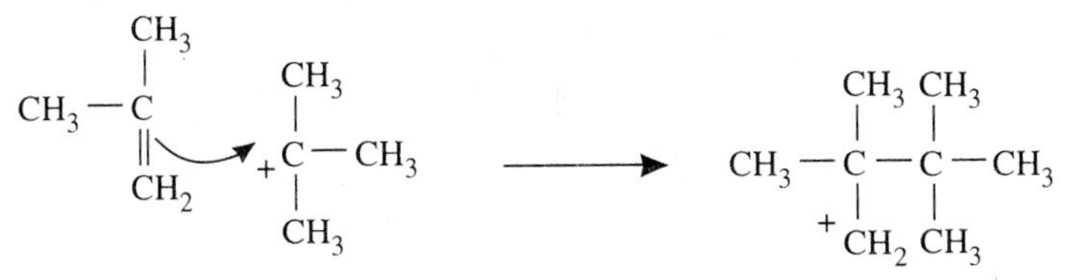

65. pi, carbon, hydrogen, neutral, positive,

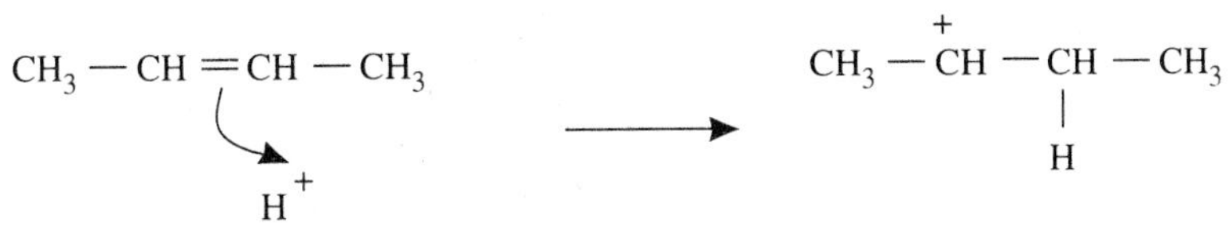

66.

The alternative,

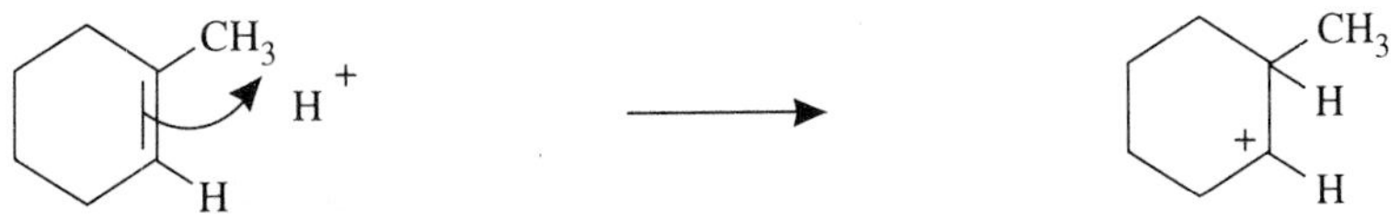

is less likely since it results in a secondary carbocation.

67. carbon, oxygen

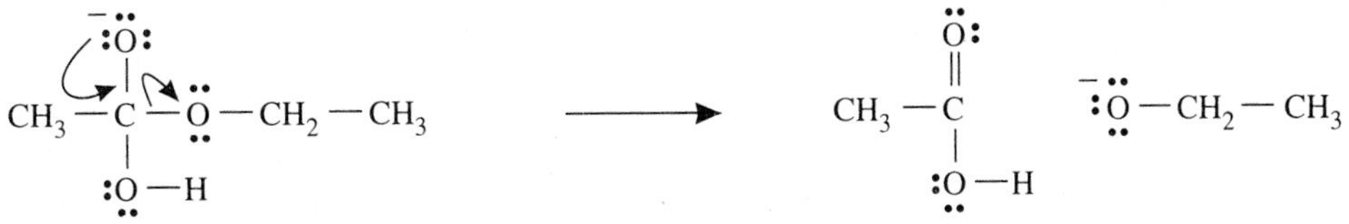

68.

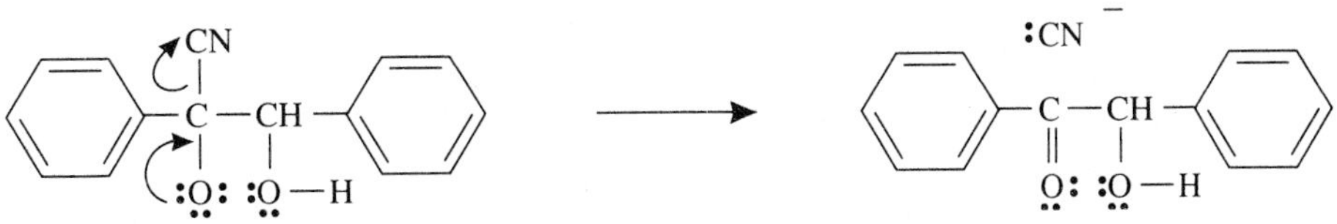

69. carbon, oxygen, positive, neutral

70. $R-C(R)(OH)-\overset{+}{O}(H)-CH_3 \longrightarrow R-C(R)=\overset{+}{O}-H \quad + \quad H-O-CH_3$

71. carbon, carbon, positive, neutral

72. carbon, carbon, carbon, positive, neutral

$$CH_3-CH(H)-\overset{+}{C}H-CH_3 \longrightarrow CH_3-CH=CH-CH_3 \quad + \quad H^+$$

73.

$$CH_2(H)-\overset{+}{C}(CH_3)-CH_2-C(CH_3)_2-CH_3 \longrightarrow CH_2=C(CH_3)-CH_2-C(CH_3)_2-CH_3 \quad + \quad H^+$$

$$CH_3-\overset{+}{C}(CH_3)-CH(H)-C(CH_3)_2-CH_3 \longrightarrow CH_3-C(CH_3)=CH-C(CH_3)_2-CH_3 \quad + \quad H^+$$

74. carbon, carbon, positive, neutral

75.

76.

77.

78. 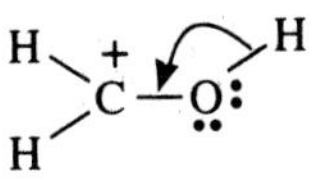 H, H, $\overset{+}{C}$—O:, H

79. 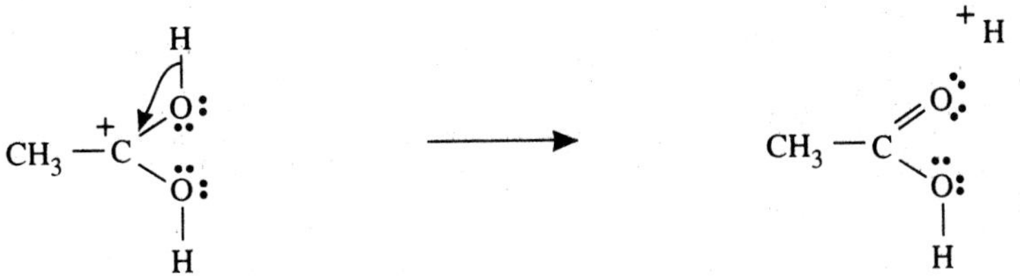 H, O:, $CH_3$—$\overset{+}{C}$, O:, H ⟶ $^+$H, O:, $CH_3$—C, O:, H

80. 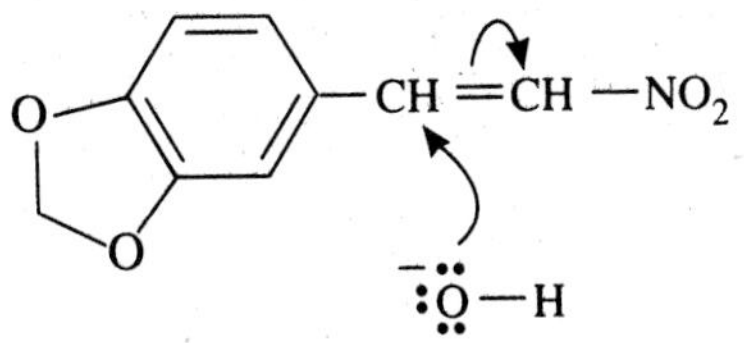 O, O, —CH=CH—$NO_2$, $^-$:O—H

81. 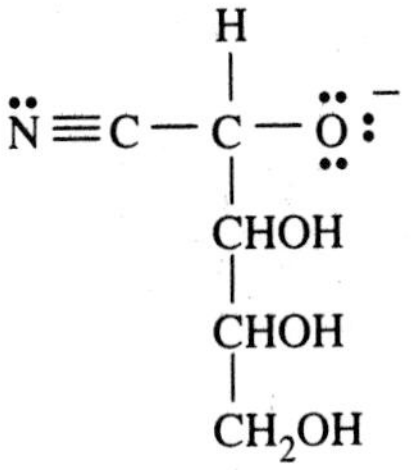 H, N≡C—C—O:$^-$, CHOH, CHOH, $CH_2OH$

82. 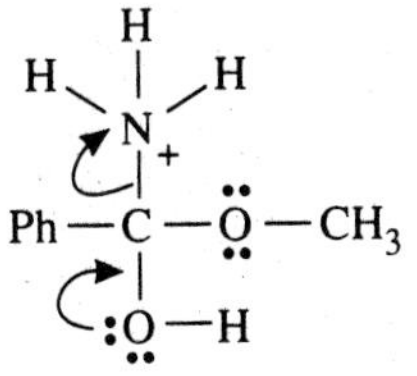 H, H, H, $N^+$, Ph—C—O—$CH_3$, :O—H

83. 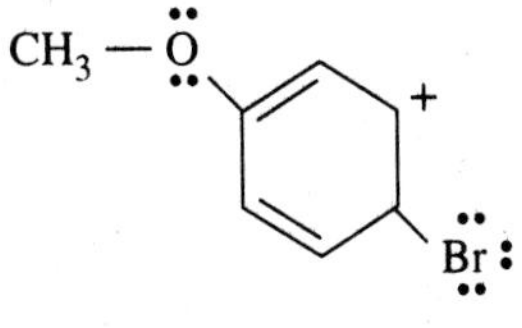 $CH_3$—O, +, Br:

84.

85.

86.

87.

88.

89. 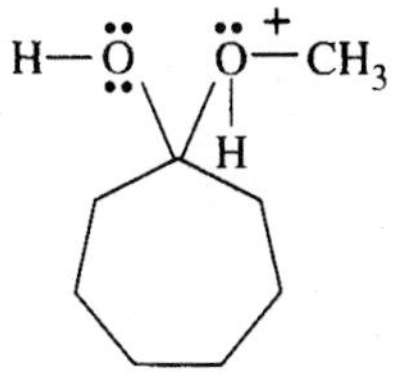 $H-\ddot{O}$ $\ddot{O}^{+}-CH_3$ H

90. 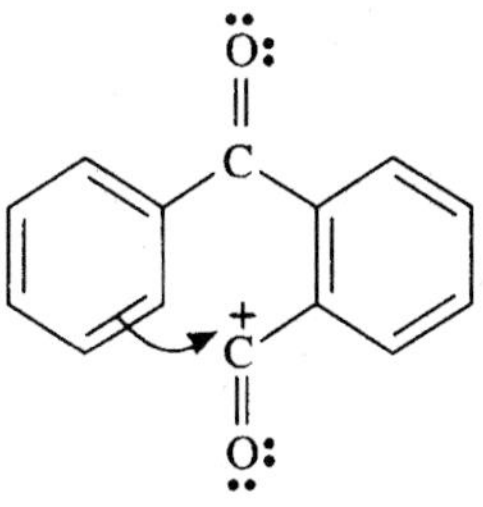 $\ddot{O}$: C $\overset{+}{C}$ $\ddot{O}$:

91. 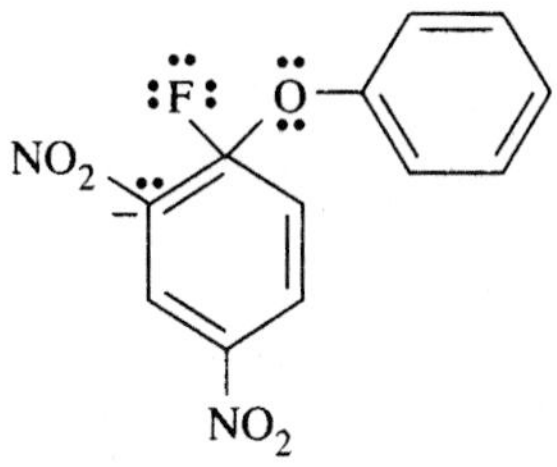 :F: O $NO_2$ $NO_2$

92. 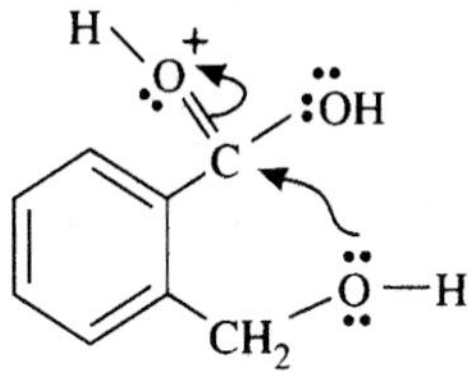 H $O^{+}$ :OH C O—H $CH_2$

93. (a) 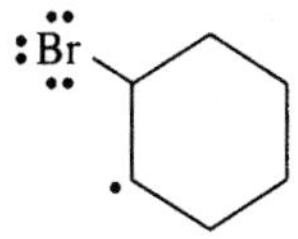 :Br

(b) 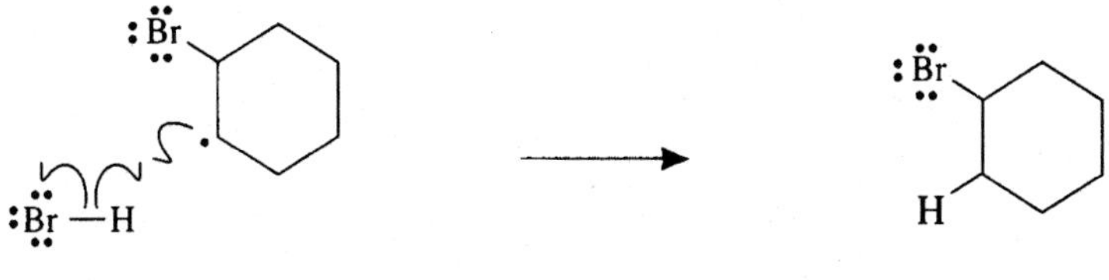 :Br :Br —H :Br H

94. (a) $CH_2{=}CH{-}CH_3$ $\cdot\ddot{Br}:$

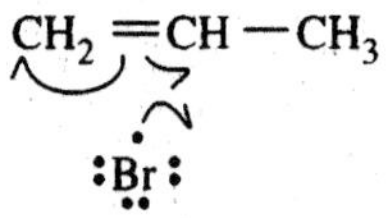

(b) $CH_2{=}CH{-}CH_3$ $\cdot\ddot{Br}:$ ⟶ $\underset{|\atop :\ddot{Br}:}{CH_2}{-}\dot{C}H{-}CH_3$

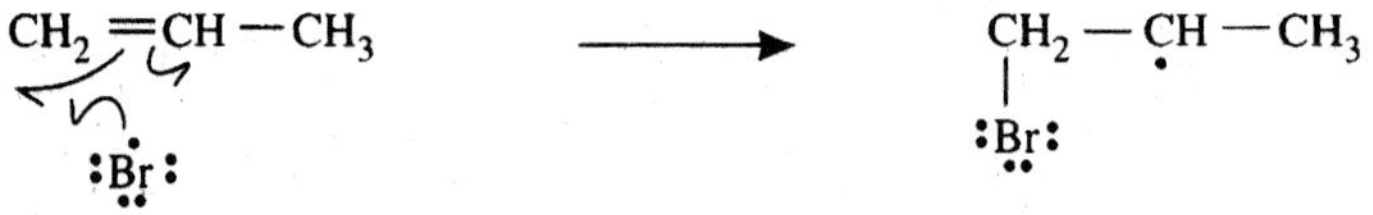

95. $Cl_3C{-}CH_2{-}\dot{C}H{-}CH_2{-}CH_2{-}CH_3$

96. $Cl_3Si\cdot$ $CH_2{=}CH{-}CH(CH_3)_2$ ⟶ $Cl_3Si{-}CH_2{-}\dot{C}H{-}CH(CH_3)_2$

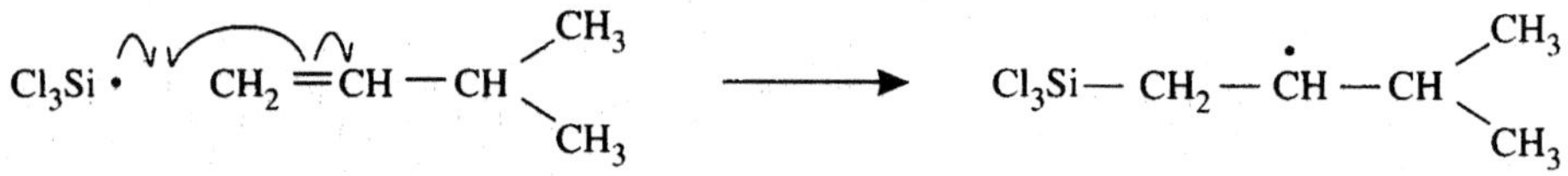

97. cyclopentyl$\cdot$ $CH_2{=}CH{-}CN$ ⟶ cyclopentyl$-CH_2{-}\dot{C}H{-}CN$

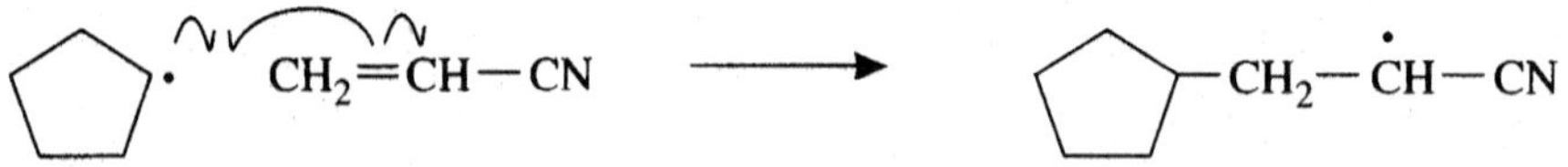

98. (a) $R'_3\,Sn\cdot$ :S: N ⟶ $R'_3{-}Sn{-}\ddot{S}$ N

:O—C—R

‖

:O

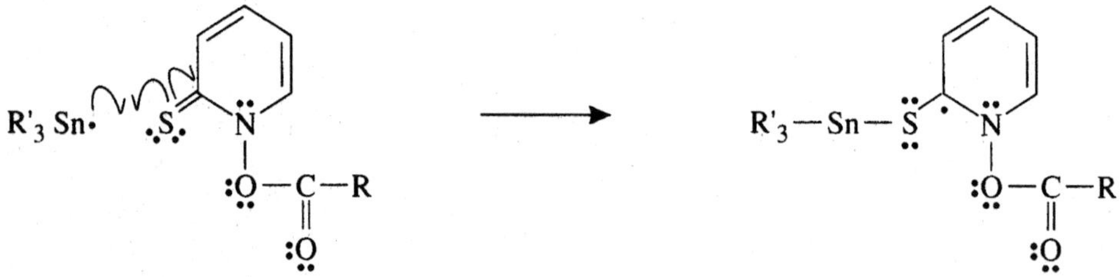

(b) $R'_3{-}Sn{-}\ddot{S}$ N

:O—C—R

‖

:O

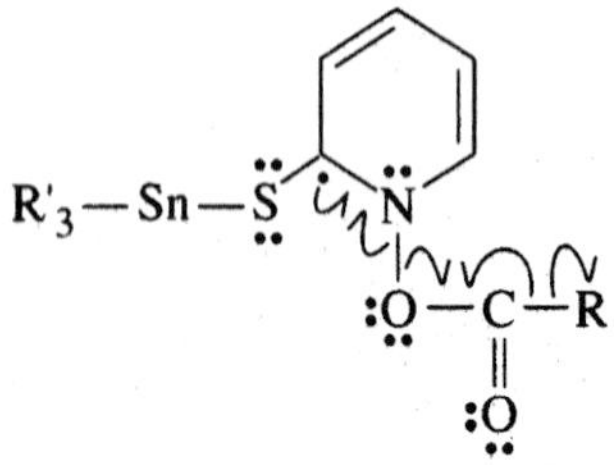

99. :Br:

$CH_3{-}CH{-}CH{-}CH_3$

H

$H{-}\ddot{O}:^-$

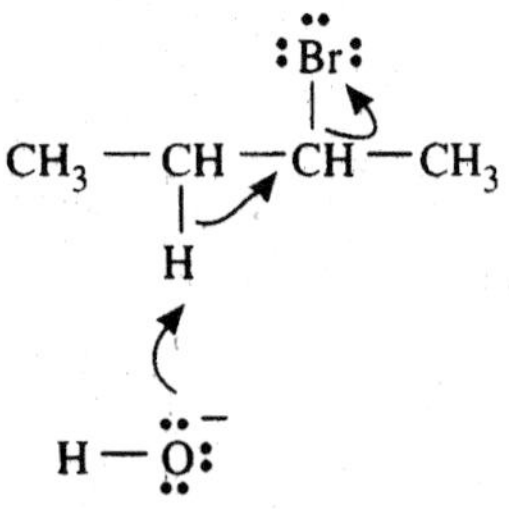

100. hydrogen, carbon, neutral, negative,

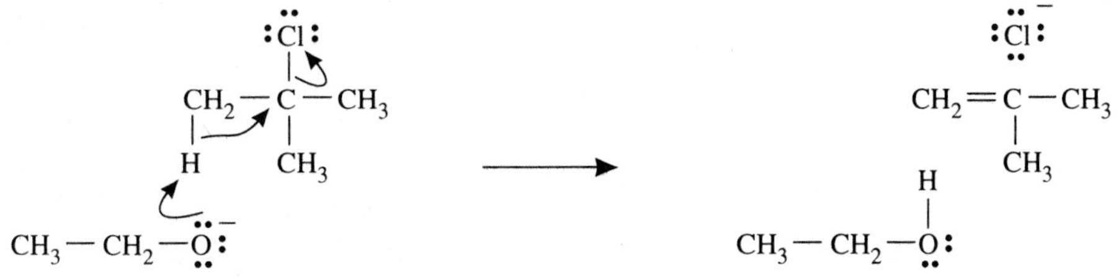

101.

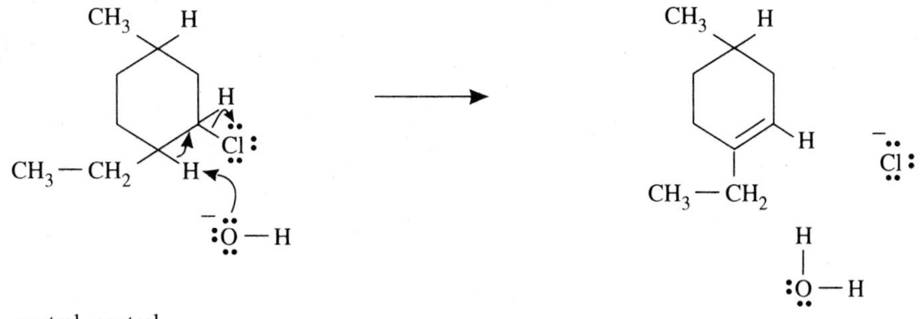

102. neutral, neutral,

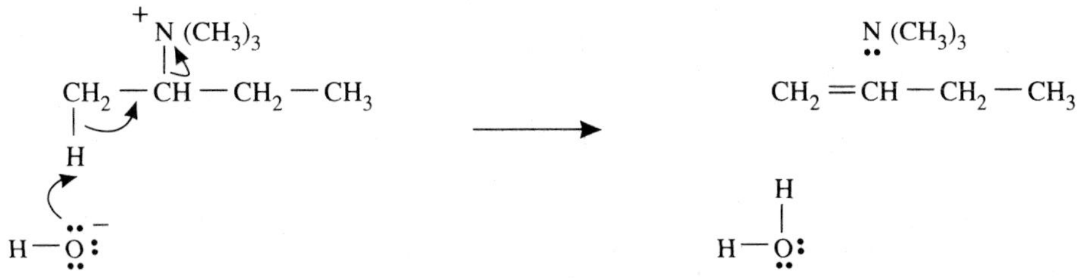

103.

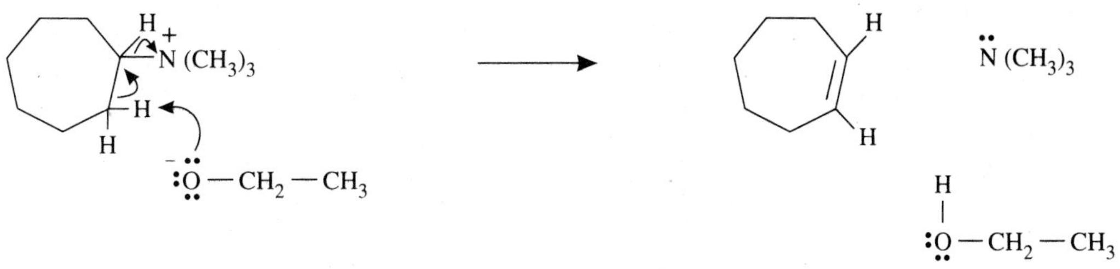

104.

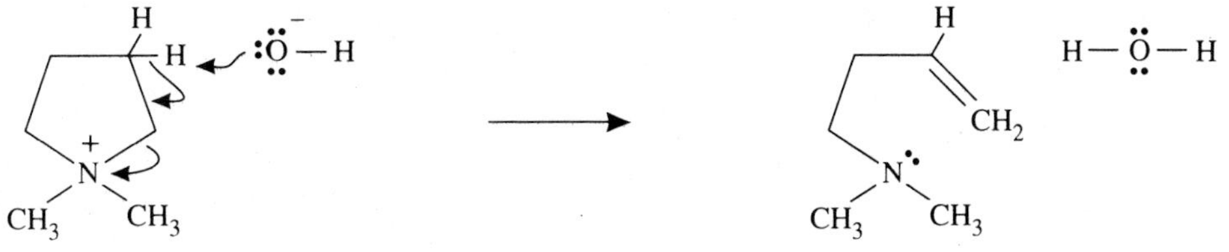

105. $\overset{1}{CH_3}-\overset{2}{CH}=\overset{3}{CH}-\overset{4}{CH_3}$ , positive, negative

$:\ddot{Cl}-\ddot{Cl}:$

106. H, H, $H-\ddot{I}:$

107.

$C_6H_5-CH=CH-C_6H_5 \quad H-\ddot{I}: \longrightarrow C_6H_5-\overset{+}{CH}-CH(H)-C_6H_5 \quad :\ddot{I}:^-$

108. $CH_3-CH(H)=\overset{+}{CH_2}$ $\quad :\ddot{Cl}:^-$

109. $CH_3-C(CH_3)=CH_2 \quad H-\ddot{I}: \longrightarrow CH_3-\overset{+}{C}(CH_3)-CH_2(H) \quad :\ddot{I}:^-$

$CH_3-C(CH_3)=CH_2 \quad H-\ddot{I}: \longrightarrow CH_3-C(CH_3)(H)-CH_2^+ \quad :\ddot{I}:^-$

110. $C_6H_5-CH=CH_2 \quad :\ddot{Br}-\ddot{Br}: \longrightarrow C_6H_5-\overset{+}{CH}-CH_2(:\ddot{Br}:) \quad :\ddot{Br}:^-$

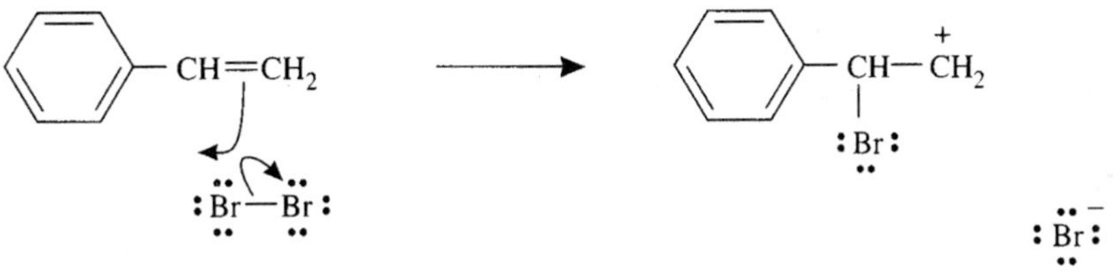

$C_6H_5{-}CH{=}CH_2 + :\ddot{Br}{-}\ddot{Br}: \longrightarrow C_6H_5{-}CH(\ddot{Br}){-}\overset{+}{C}H_2 \quad :\ddot{Br}:^-$

111. $CH_3{-}C(\ddot{O}H){=}\ddot{O}$ $\quad (CH_3)_3C^+$

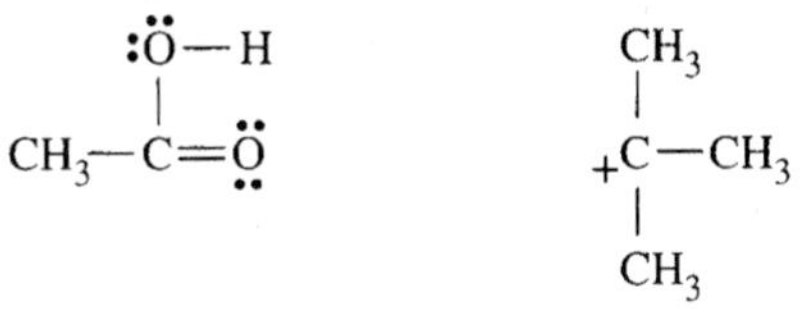

112. $H_2C{-}H$, $CH_3{-}C^+(CH_3)$ ; $H_2\ddot{O}$

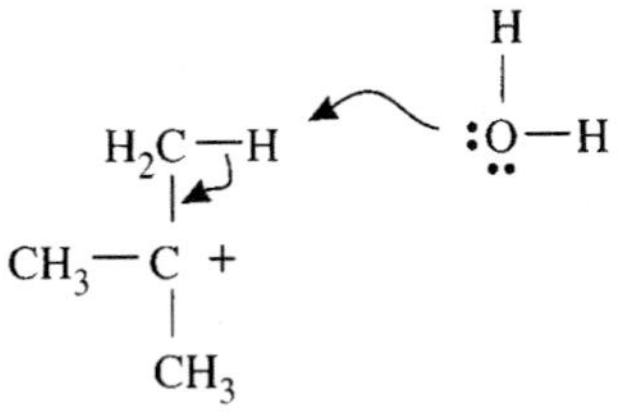

113. $H_3O^+$

$Ph{-}C({=}O){-}\ddot{N}H{-}Ph$

$H_2\ddot{O}$

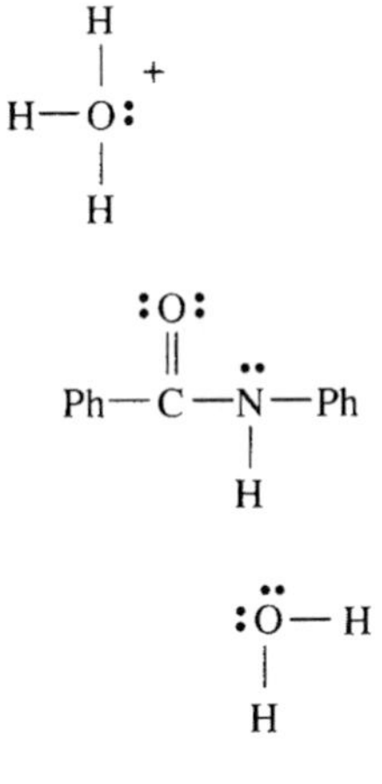

114. $Ph{-}C(\ddot{O}H)(H){-}O^-$ ; $Ph{-}C(H){=}O$

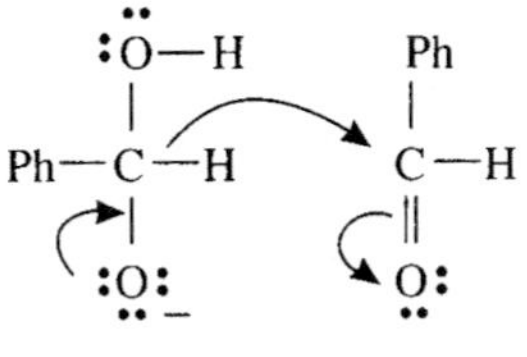

115. $C_6H_5NO_2$ (H ortho to $NO_2$) ; $H{-}\ddot{O}{-}SO_3H$

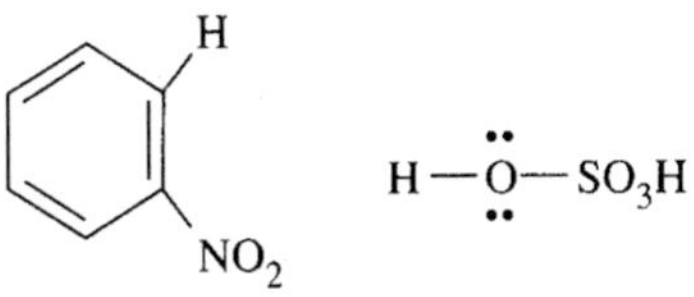

116.

$$\mathrm{Ph{-}CH_2{-}C(OH)(OH){-}\overset{+}{N}H_3} \quad {}^{-}\mathrm{O{-}H} \quad \mathrm{H{-}O{-}H}$$

117.

$$\mathrm{CH_3{-}C(OH)(OH){-}\overset{+}{O}(H){-}CH_2{-}CH_3} \quad \mathrm{H_2O} \quad \mathrm{H_3O^+}$$

118.

$$\mathrm{CH_3{-}C(CH_3)(H){-}\overset{+}{C}H_2} \longrightarrow \mathrm{CH_3{-}\overset{+}{C}(CH_3){-}CH_2{-}H}$$

119.

$$\mathrm{CH_3{-}\overset{+}{C}(CH_3){-}CH(CH_3){-}CH_3}$$

120.

$$\mathrm{CH_3{-}C(CH_3)(CH_3){-}\overset{+}{C}H_2} \longrightarrow \mathrm{CH_3{-}\overset{+}{C}(CH_3){-}CH_2{-}CH_3}$$

121.

$$\mathrm{H{-}C(H)(CH_3){-}\overset{+}{C}H_2} \longrightarrow \mathrm{H{-}\overset{+}{C}(CH_3){-}CH_2{-}H}$$

122. $CH_3-CH_2-C(CH_3)(H)-\overset{+}{C}H_2 \longrightarrow CH_3-CH_2-\overset{+}{C}(CH_3)-CH_2-H$

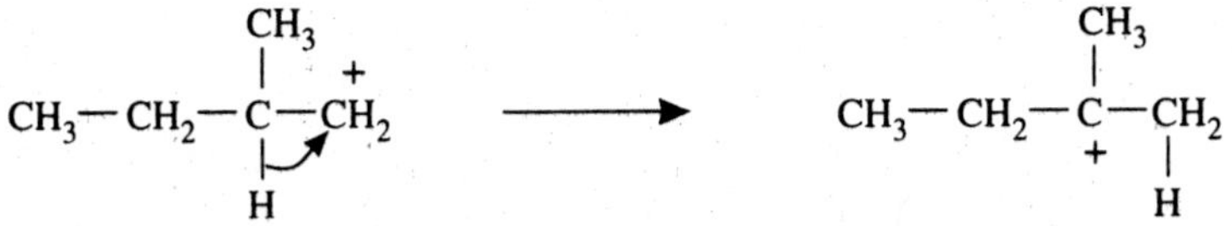

123. $CH_3-C(CH_3)(H)-\overset{+}{C}H-CH_2-H \longrightarrow CH_3-\overset{+}{C}(CH_3)-CH(H)-CH_2-H$

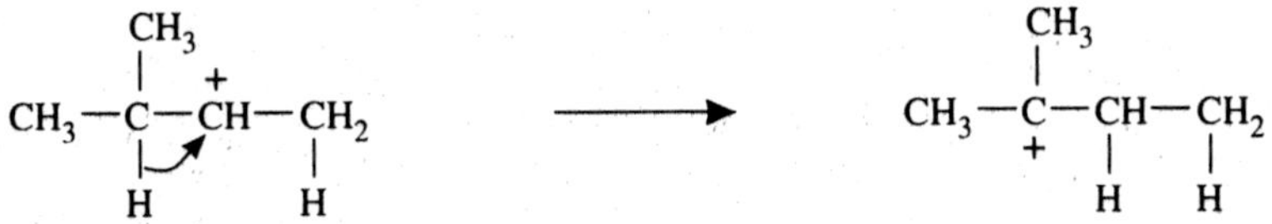

124. $CH_3-C(CH_3)(\ddot{O}H)-\overset{+}{C}(CH_3)-CH_3 \longrightarrow CH_3-\overset{+}{C}(\ddot{O}H)-C(CH_3)(CH_3)-CH_3$

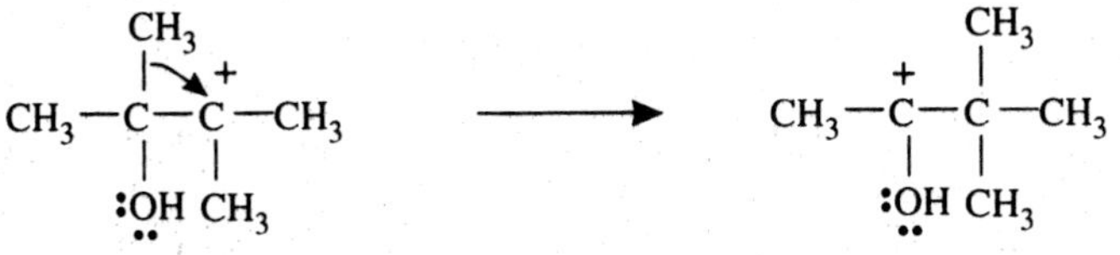

125. $CH_3$, $CH_3$, +, H $\longrightarrow$ $CH_3$, +, $CH_3$, H

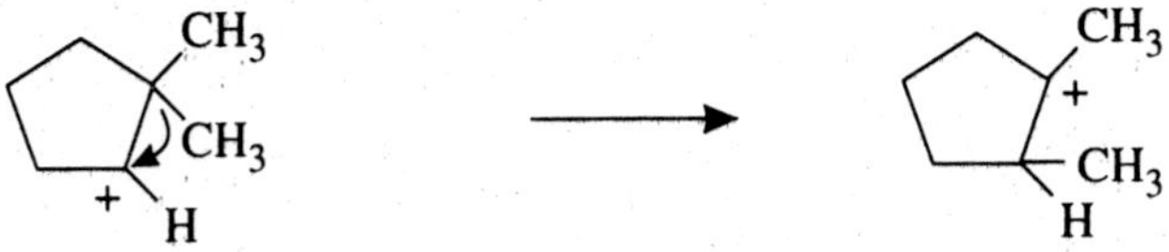

126. Ph—C—CH—, Ph

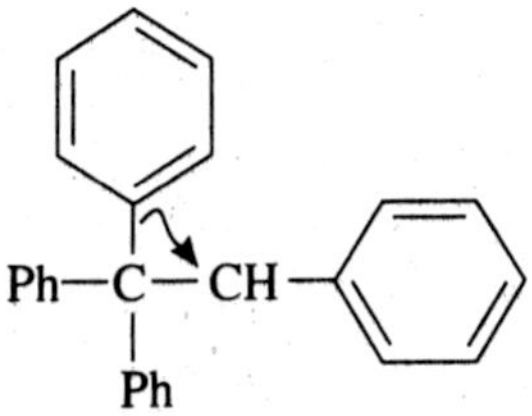

127. Ph, +, H $\longrightarrow$ Ph, +, H

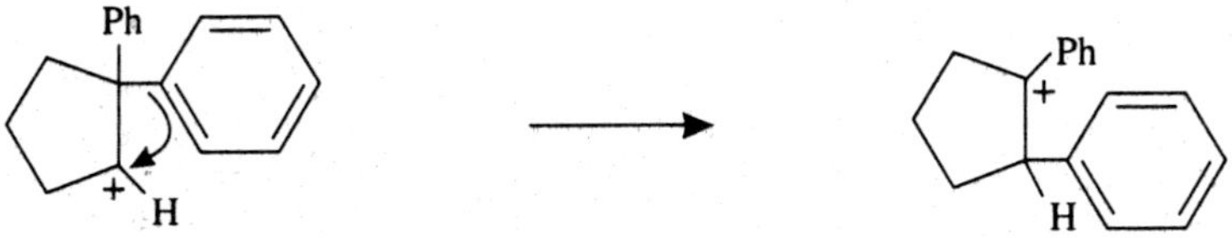

28. Ph—$\overset{+}{C}$——C—$\ddot{O}H$ $\longrightarrow$ Ph—C——$\overset{+}{C}$—$\ddot{O}H$

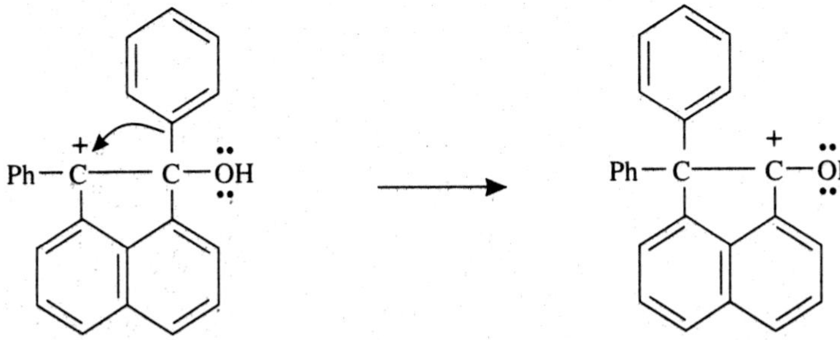

129.

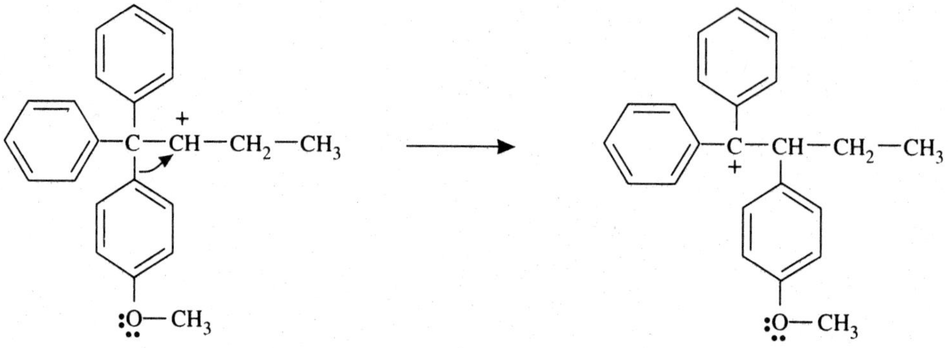

130.

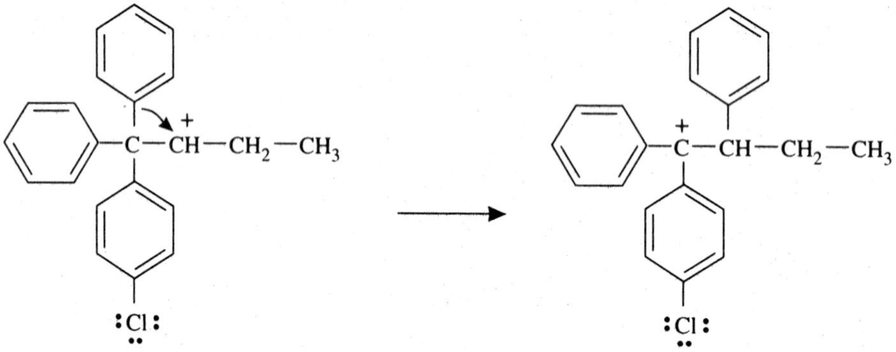

131.

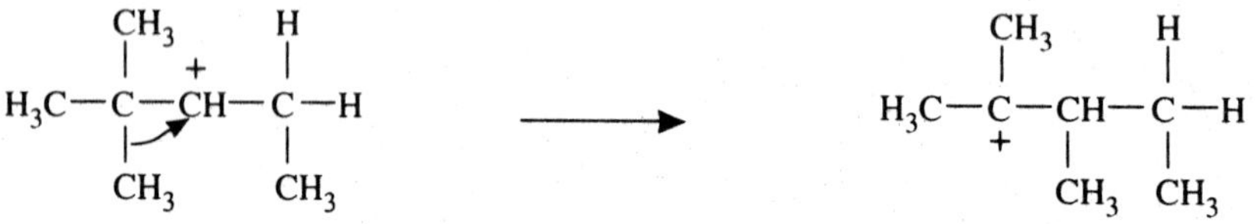

132.

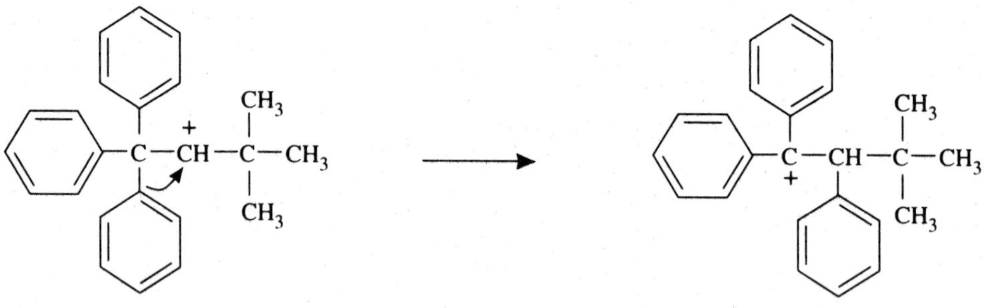

133.

134.

135.

136.

137.

138.

139.

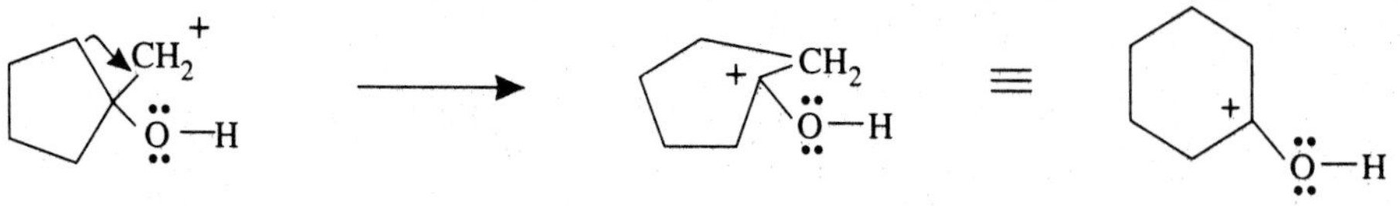

140.

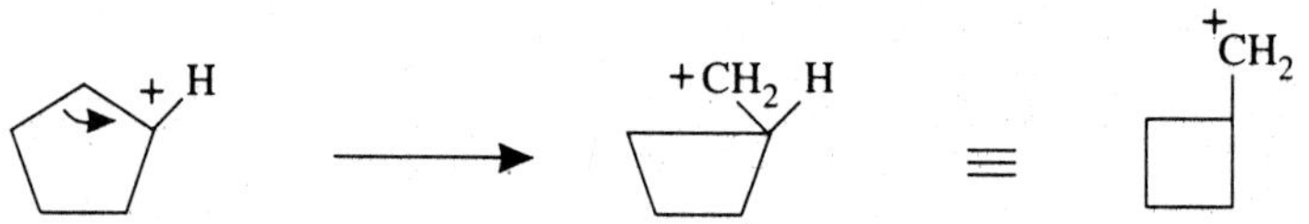

141.

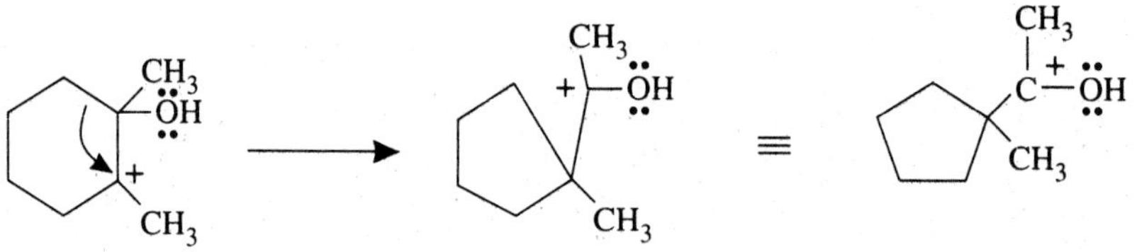

142.

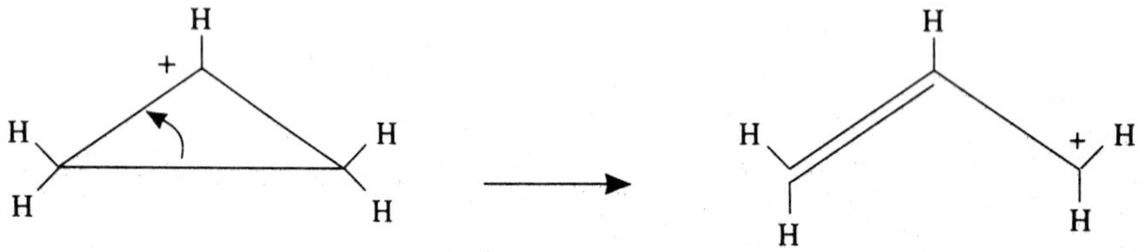

143.

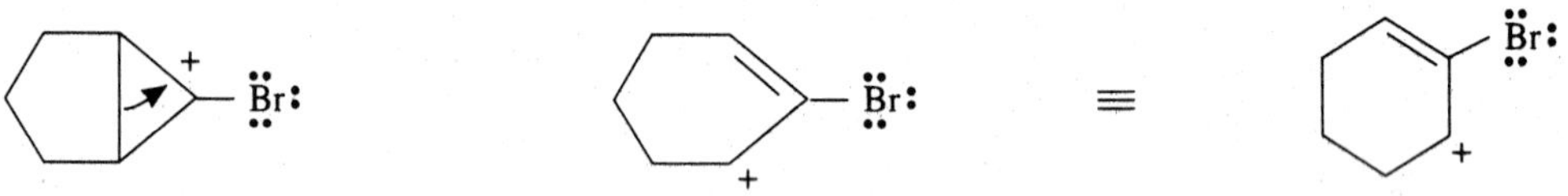

144.

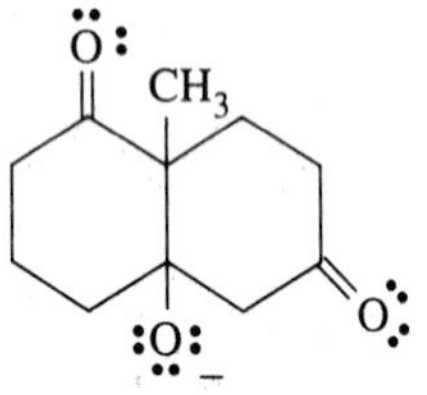

145. $CH_2{=}CH{-}CH_2{-}CH_2{-}CH_2{-}\dot{C}H_2$

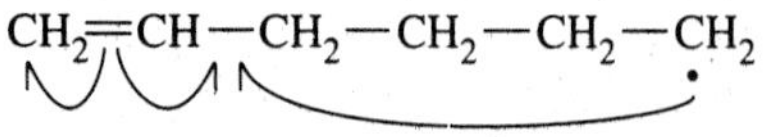

146.

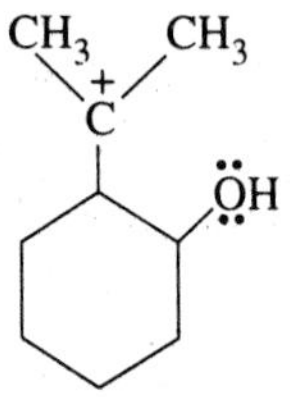

147. $:\ddot{\text{O}}^{-}—CH_2—CH_2—CH_2—CH_2—\ddot{\text{Br}}:$

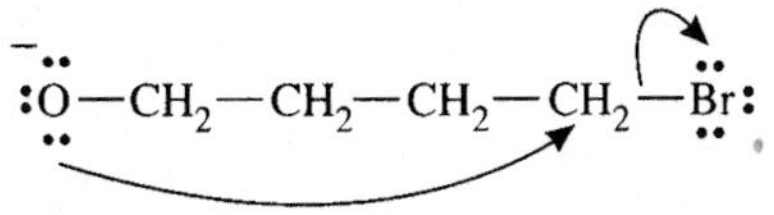

148. (cyclopentane with two COOR groups) + $:\ddot{\text{Cl}}:^{-}$

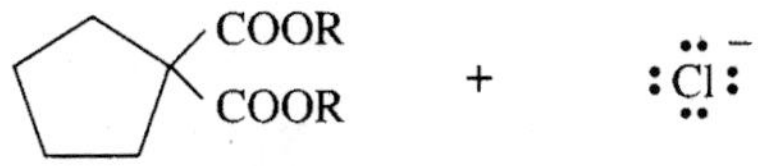

149. $CH_2{=}CH—C(=\ddot{\text{O}})—\ddot{\text{O}}—(CH_2)_9—\dot{C}H_2$

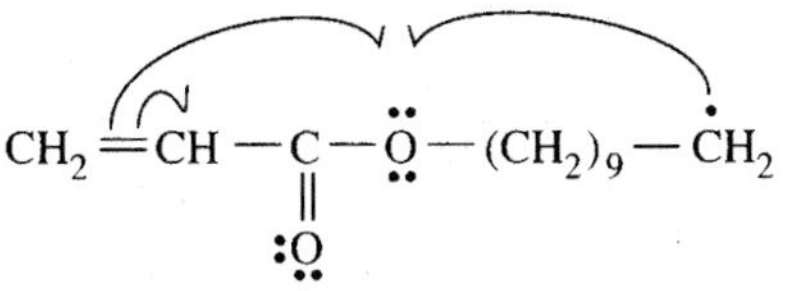

150. (2,2-dimethylcyclopentyl)$C^{+}(CH_3)_2$

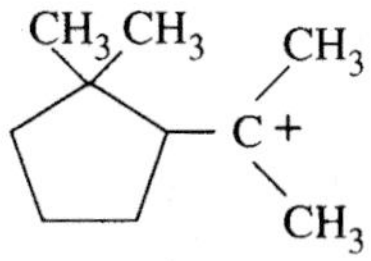

151.

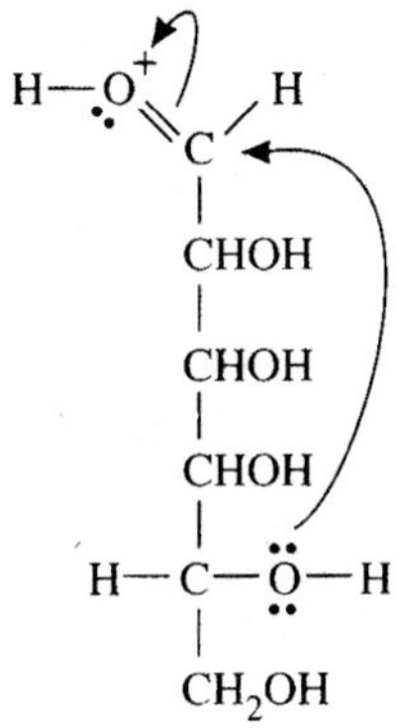

$H—\ddot{O}^{+}{=}C(H)—CHOH—CHOH—CHOH—CH(\ddot{O}H)—CH_2OH$

152. (cyclohexyl)$\dot{C}H_2$

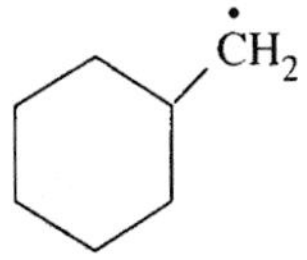

153. $R—\ddot{O}—C(=\ddot{O})—\ddot{C}H^{-}—CH_2—CH_2—CH_2—C(=\ddot{O})—\ddot{O}—R$

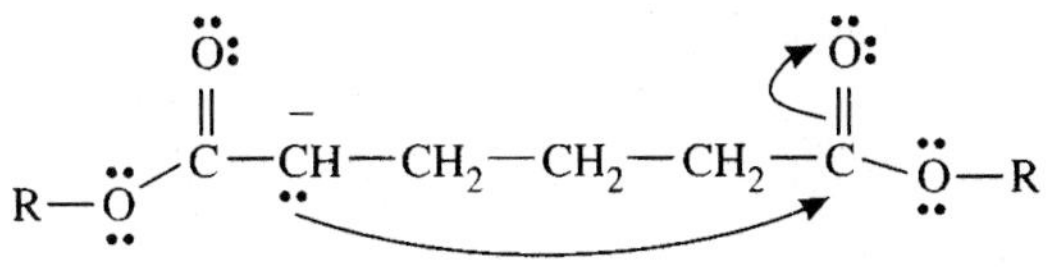

# 4

# On Solving Mechanism Problems

**T**his chapter will provide some approaches to solving those troublesome problems that state, in one form or another, "using curved arrow notation, provide a mechanism for the following transformation." The chapter describes a way of thinking about these problems, offers some wisdom in dealing with them, and warns of some **pitfalls to avoid, but, there is no set formula for solving these problems.** You must be prepared to think them through.

Mechanism problems, as treated here, come in various disguises, but all of them provide three items: the starting materials, the reagents/conditions, and the products. That is, you are informed of all the compounds in the pot. It is your task to describe **what the molecules do.** The single most common mistake that my students commit is to confuse mechanism problems with synthesis problems. In a synthesis problem you are given the starting material and the product and you are asked to convert—on paper—the former into the latter. In that case, your task would be to describe **what the chemist does.** In solving a mechanism problem, avoid the pit of getting halfway through the problem, not seeing immediately how the molecules will proceed further, and—in panic—supplying a reagent that was not given. You will end up with a schizophrenic solution that is not likely to earn many points.

The authors of most texts introduce mechanism problems gently. At first, such problems are simple variations on a mechanism you have just seen. For example, if you have just studied the electrophilic addition of HCl to a double bond, a similar addition of HBr will proceed *via* the same process. Repetitive drill of this sort is preparing you for the more challenging problems. The tougher problems come soon enough, but you cannot solve them without a strong foundation in the central mechanisms of the course.

So what are the central mechanisms of the course and how many are there? Well, it depends on who is counting; but these are my eight, and if you come to the arena with them firmly in hand, you will be able to wrestle with most problems successfully. I promise.

## Weeks's List of Eight Central Mechanisms for First-Year Organic Chemistry

Electrophilic addition to alkenes and alkynes

Nucleophilic aliphatic substitution-elimination

Electrophilic aromatic substitution

Nucleophilic addition to carbonyl and similar groups

Nucleophilic acyl substitution

Substitutions and condensations on a carbon atom α to a carbonyl

Free radical substitutions

Free radical additions

Harder problems appear as each new mechanism is introduced. Near the end of the course you will need all the central mechanisms. On the other hand, having to know fewer than ten mechanisms isn't so bad. It will help to have a good feeling for organic structure, and it can't hurt to know your reactions. In other words, the more organic you know, the easier this is going to be.

Before I stop preaching, I must relate an anecdote. This really happened, trust me. A few years ago I was teaching the third quarter of organic and I noticed that there was a senior on the class roster. It was a big class and he kept to himself, but he did well and earned an "A." During senior week he stopped by to pick up his final exam and we talked for a bit. I asked him why he took the course as a third-quarter senior and, as I had guessed, he had a provisional admission to medical school, the provision being that he take and pass the third quarter of organic. Then I asked him what I really wanted to know, namely, was it difficult to come back after being away from the course for two years? He said, "Well, of course I had to go back and review a few things but actually it was much easier. When I took the first two quarters as a sophomore I did very well in the course, but my friends and I memorized *every mechanism in the book!*" (his emphasis) "But in the first week of your course I realized that *it's just nucleophiles and electrophiles.*" (my emphasis)

That is a great truth, simply stated. Like all great truths it needs to be developed, and like most simple statements it's not entirely true. But when I heard this guy, I thought, Wow! If I could replay that for all my students, it would save them a lot of time and frustration.

*     *     *     *     *

I shall present a series of problems in roughly the order in which the relevant mechanisms are presented in most texts. Along the way I shall emphasize strategies, marked with bullets •, that I have found helpful. I begin with the *Prime Directive* and a nod to you Trekkies.

- Don't begin by scribbling chemistry. Instead, look at the big picture. Try to discern the relationship between the starting material(s) and the product(s) and ask, "What went on here?"

1. A simple example. Propose a mechanism for the decomposition of this anion.

$$^{-}\!:\ddot{\underset{..}{O}}-C(=\ddot{O}:)-\ddot{\underset{..}{O}}:-C(CH_3)_3 \xrightarrow{\text{spontaneous}} :O=C=\ddot{O}: \;+\; ^{-}\!:\ddot{\underset{..}{O}}-C(CH_3)_3$$

You probably see this right away, but humor me: Ask yourself what has happened here. One starting material has become two products, so at least one bond had to break. Which one? The landmark is the *tert*-butyl group with an —O⁻ attached. If you break that off the starting material (no chemistry yet), a carbon attached to two oxygens remains. Isn't that comforting? Now you can try some chemistry, knowing which bond you want to break and remembering that you should push electrons *away* from a negative charge. Try it on the structure above.

This first exercise is trivial, but it makes the point that the journey is easier if you know where you're going. Is that an ancient Asian proverb?

2. Another decarboxylation.

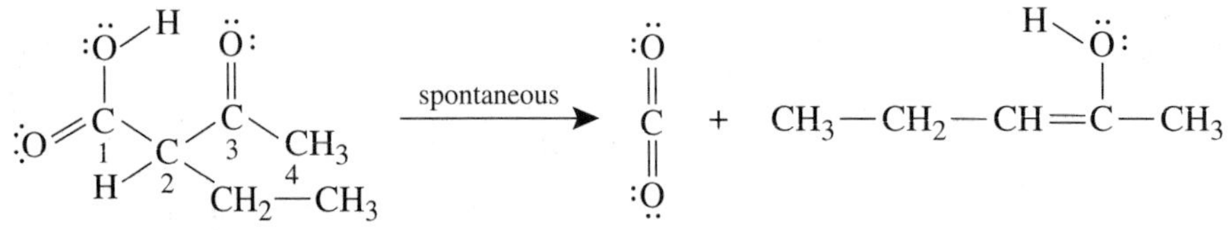

$$HO{-}\overset{1}{C}(=O){-}\overset{2}{C}H(\overset{4}{C}H_2{-}CH_3){-}\overset{3}{C}(=O){-}CH_3 \xrightarrow{\text{spontaneous}} O{=}C{=}O + CH_3{-}CH_2{-}CH{=}C(OH){-}CH_3$$

You will not begin by __________ chemistry. You will first ask, ________________ . One starting material has become _____ products. Taking the product, $CO_2$, as a landmark, the only carbon in the starting material that has two oxygens attached to it is C. If you break the C1—C2 bond, the other fragment will have _____ carbon atoms. Might you be on the right track? ______ (Y/N). But, there is a carboxyl hydrogen atom that will end up in the wrong place. What to do? If it could just be moved over to the oxygen on C3 we would be in business. Let's trying pushing electrons to (1) break C1—C2, (2) form a double bond between C2 and C3, (3) transfer a hydrogen from the carboxyl group to the C3 carbonyl and, (4) end up with another C=O on the emerging $CO_2$ molecule. Go ahead.

3. Starting with the enol product of the previous problem, propose a reasonable mechanism for this "spontaneous" reaction.

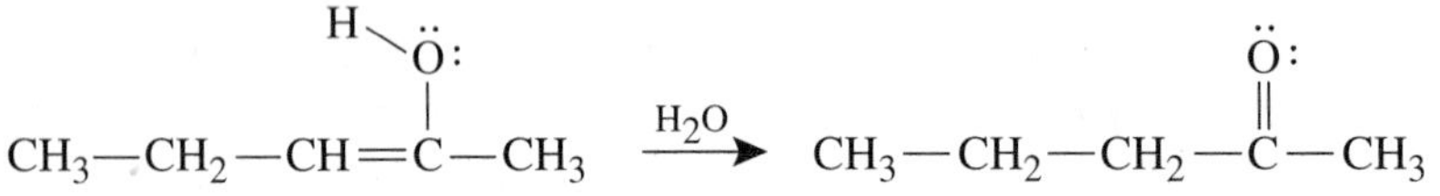

$$CH_3{-}CH_2{-}CH{=}C(OH){-}CH_3 \xrightarrow{H_2O} CH_3{-}CH_2{-}CH_2{-}C(=O){-}CH_3$$

One starting material, $C_5H_{10}O$, with a continuous chain of _____ carbon atoms, has become one product with the same molecular formula and the same continuous chain. The double bond has moved from between two carbons to between a carbon and a(n) ________ and a hydrogen has moved from oxygen to _______ . Watch out! I've set a trap. Using the previous exercise as a model, it is very tempting to do this.

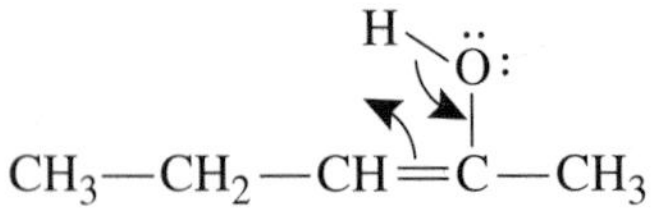

$$CH_3{-}CH_2{-}CH{=}C(OH){-}CH_3$$

- Try not to propose high-energy transition states or intermediates.

If you had an uncomfortable feeling about a four-membered transition state, you were right. In exercise 2 the transition state was a six-membered ring, and we know that they are good. Four-membered rings are bad, too much angle strain.

- Be prepared to reject an approach if it's not working (I'll come back to this).

How are we going to move that proton? Instead of moving a single proton from oxygen to carbon, why not use the solvent to take one proton off oxygen and put another on carbon? Try it.

$$H_2O + CH_3-CH_2-CH=C(OH)-CH_3 + H_2O \longrightarrow H_3O^+ + CH_3-CH_2-CH(H)-C(=O)-CH_3 + {}^-OH$$

______ arrows

The hydroxide and hydronium ions will instantly neutralize each other. "Spontaneous" used in this sense is taken to mean that the reaction will proceed under the stated conditions, using the thermal energy available at ambient temperature. The solvent is always part of the conditions.

- Do not add reagents.

When the difficulty of the four-centered transition state became apparent, it would have been *so* easy to throw some acid into the pot, but no acid was stipulated in the problem! As an aside, this reaction goes quite slowly at neutrality and is greatly accelerated by acid catalysis.

All three problems so far are one-step mechanisms, and you have been doing this sort of thing all through Chapter 3. Here are some mental steps you should take.

- Look at the big picture.
- Decide where you're going.
- Only then begin to think about the chemistry.
- Make your mechanism reasonable according to the principles you have learned.
- Reject an approach if it's not working. Don't get married to a single idea.
- Don't add reagents.

* * * * *

The next section will be based on two mechanisms usually presented early in the course:

electrophilic addition to alkenes and nucleophilic substitution-elimination ($S_N1$-$S_N2$/E1-E2). It will include several themes that keep turning up in these problems.

## Theme I—The Intercepted Intermediate

4. Propose a mechanism for this reaction.

$CH_3$ $\xrightarrow[NaCl_{excess}]{Br_2}$ $CH_3$ Cl: Br: + NaBr

The big picture: An alkene has lost its double bond. Halogen atoms have appeared at the termini. This looks like electrophilic addition to the alkene, which is one of those mechanisms you must know. But when $Br_2$ is used, bromine atoms appear at each end of the erstwhile double bond. The only source of chlorine is NaCl.

With that background, write the first step in the mechanism of bromine's addition to the double bond. (If you can't, you're not ready for this chapter).

Now comes the intercepted intermediate part. Clearly, finishing off this reaction in the usual way will lead to the 1,2-dibromo compound. But, instead of using $Br^-$ as the nucleophile, try using $Cl^-$.

You have completed the task. There are some other factors that could be considered. First, the less nucleophilic chloride ion competes successfully because it is present in excess. Second, the regiochemistry is correct for opening the bromonium ion and, third, the stereochemistry is anti.

But you needn't worry about these points. You were given the product, and no stereochemistry was indicated in its structural formula. Your task was to get to the product stipulated in the problem.

- Don't answer questions you are not asked.

That last point is a bit flippant. If you encounter this problem while you are studying, of course you should think about these other factors. But during an exam, where time is your most precious commodity, answer only the questions you are asked.

5. Propose a reasonable mechanism for this reaction, using curved arrow notation.

$$\underset{6}{CH_2}{=}\underset{5}{CH}{-}\underset{4}{CH_2}{-}\underset{3}{CH_2}{-}\underset{2}{CH_2}{-}\underset{1}{CH_2}{-}\ddot{\underset{..}{O}}{-}H \xrightarrow[H_2O]{H_2SO_4}$$

CH$_3$ :O:

What is going on here? Where did that ring come from? There are six carbons and one oxygen in both starting material and product. When a ring forms from a single molecule, one end must bite the other. Taking the oxygen as a landmark, it will have to become attached to C ________ in order for a six-membered ring to form. That's comforting because the —OH and the C=C are the functional groups and you expect the chemistry to happen there. Another comforting observation: The four connected —$CH_2$— groups appear in both starting material and product. So the $CH_3$ in the product comes from C in the starting material ________.

Enough analysis. Let's try some chemistry. One could protonate the hydroxyl and generate a carbocation. It is quite a reasonable thing to try, but at some point you would see that the oxygen you need in the product is gone. You must reject this good idea because it is not working. What else could become protonated? The double bond? Try it.

(a)

If you have produced the more stable, secondary carbocation, you should see a way to close the ring so that the oxygen becomes attached to C5. Once again, an intermediate cation has been intercepted, this time by a nucleophile at the other end of the molecule.

(b)

Now it's just a matter of getting rid of that proton. I'll leave that to you.

## Theme II—Rearrangement

6. Propose a mechanism for this reaction.

$$CH_3-\underset{CH_3}{\overset{CH_3}{C}}-CH_2-\ddot{O}-H \xrightarrow{H_2SO_4} (CH_3)_2C{=}C(H)CH_3 + H_2O$$

In the presence of a strong acid, an alcohol has become an _______ plus water. Sounds like dehydration. Both the starting material and product contain _____ carbon atoms, but the longest chain in the starting material is _______; the longest chain in the product is _______. It must be a rearrangement! In the first course in organic, carbocations are almost always responsible for rearrangements. In two steps, generate a carbocation.

(a)

Now do the rearrangement part.

(b)

And the loss of a proton completes this E1 reaction.

(c)

## Theme III—An Unexpected Product

7. Propose a mechanism for this reaction.

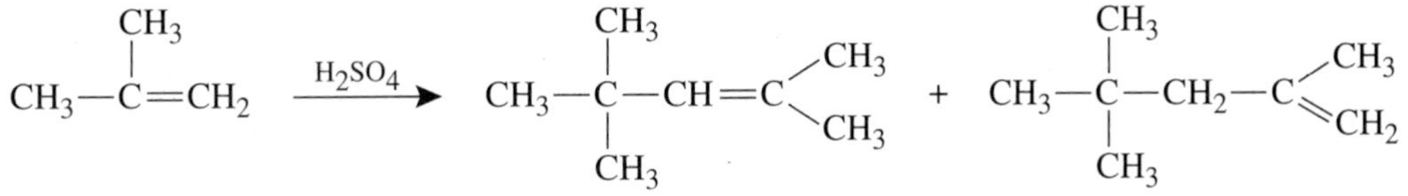

$$CH_3-\underset{}{\overset{CH_3}{\overset{|}{C}}}=CH_2 \xrightarrow{H_2SO_4} CH_3-\underset{CH_3}{\underset{|}{\overset{CH_3}{\overset{|}{C}}}}-CH=C\begin{matrix}CH_3\\CH_3\end{matrix} + CH_3-\underset{CH_3}{\underset{|}{\overset{CH_3}{\overset{|}{C}}}}-CH_2-C\begin{matrix}CH_3\\ \| CH_2\end{matrix}$$

major                minor

What's going on? A starting material with ______ carbons becomes a product with ______. There are two, isomeric products. Further observation reveals two units of the starting material in the product. Circle them in the major product above.

- Keep it simple, concentrate on one thing at a time.

Don't let two products worry you; let's do some chemistry. The nucleophilic double bond and the strong acid suggest protonation. This time, think about which carbon accepts the proton and which becomes the carbocation.

(a)

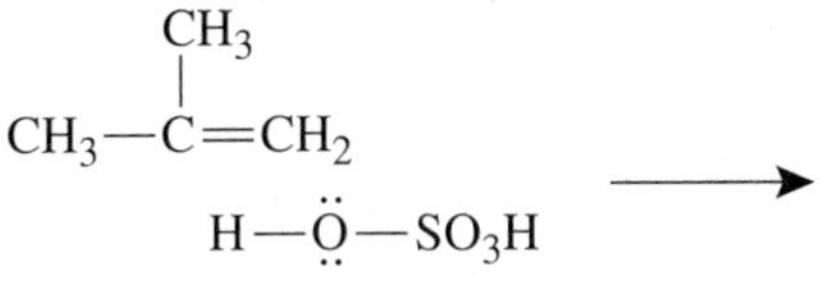

$$CH_3-\overset{CH_3}{\overset{|}{C}}=CH_2 \qquad H-\ddot{\underset{\cdot\cdot}{O}}-SO_3H \longrightarrow$$

______ supply arrows        ______ products

The carbocation is electrophilic and can attack another molecule of the starting material. Again, choose to generate the more stable cation.

(b)

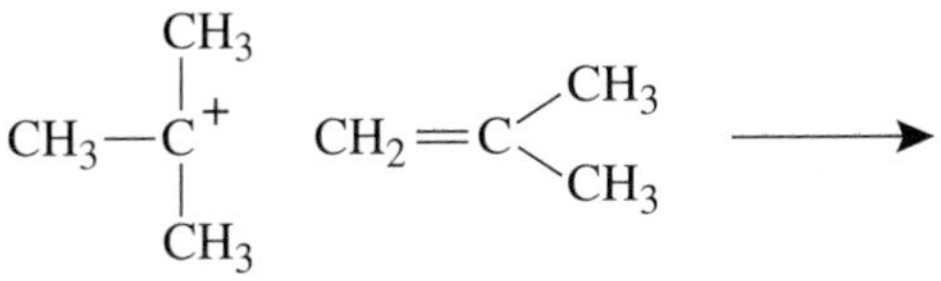

$$CH_3-\underset{CH_3}{\underset{|}{\overset{CH_3}{\overset{|}{C^+}}}} \qquad CH_2=C\begin{matrix}CH_3\\CH_3\end{matrix} \longrightarrow$$

______ supply arrows        ______ products

Doing the chemistry without worrying about the two products has worked! You ought to be able to arrive at both products from the single carbocation.

## Theme IV—Stereochemistry

8. The two reactions below are carried out under identical conditions but yield different products. Give mechanisms that account for formation of the products. Here is a problem in which you will have to grapple with the stereochemistry. You will always know this when the structures and/or names include stereochemical designations.

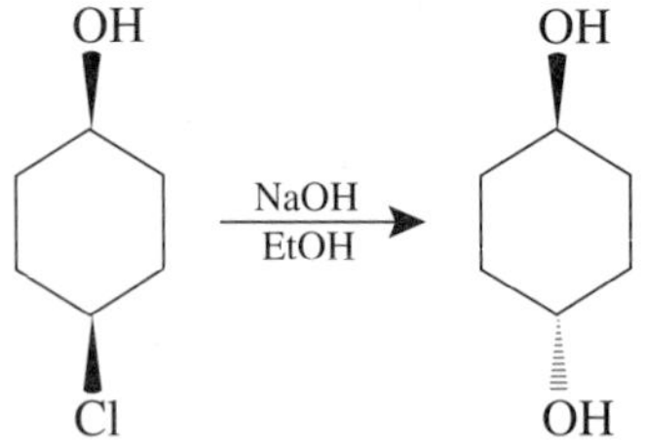

OH NaOH/EtOH → OH

Cl OH

(1)

*cis*-4-chlorocyclohexanol

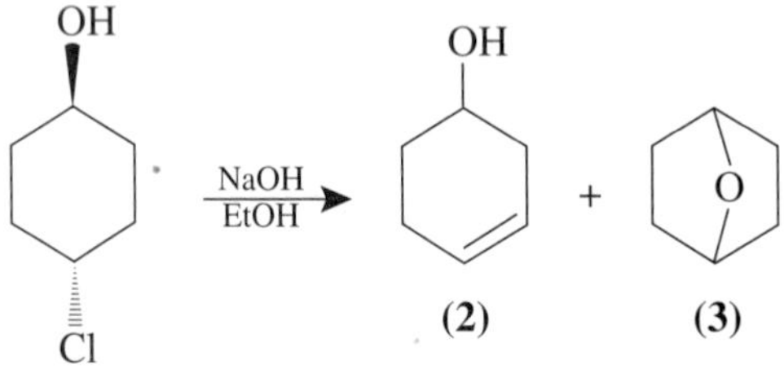

OH NaOH/EtOH → OH + O

Cl (2) (3)

*trans*-4-chlorocyclohexanol

- Keep it simple. Do one thing at a time.

First, determine the simple structural chemistry. Then address the stereochemistry and then, the unexpected product (3).

The simple chemistry. In the first reaction –OH substitutes for –Cl to form (1). In the second reaction we see the α, β elimination of HCl to form the alkene (2). Sodium hydroxide in ethanol are classic $S_N2$-E2 conditions. This sounds like familiar territory. If the first reaction is $S_N2$, the configuration will be ________________ (inverted/retained/racemized). If the second reaction is E2, the elimination will be ____________ (syn/anti).

The person who crafted this problem has added a task by the way in which the problem is presented. Although the structures designate the stereochemistry unambiguously, they are not sufficient for showing mechanisms. You must rewrite them in order to show the three-dimensional shape of the six-membered ring. I have done that for you.

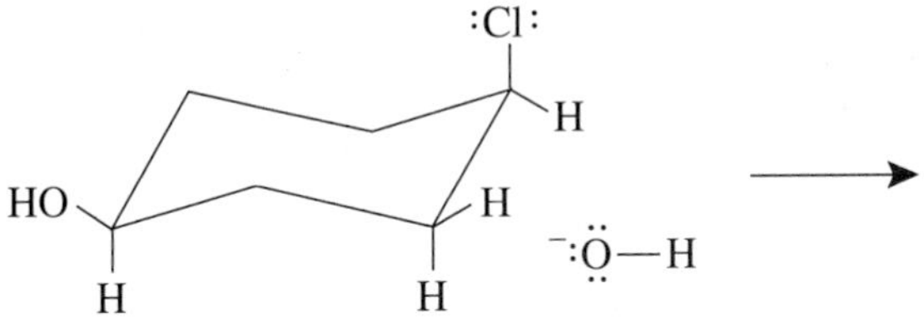

:Cl: H HO H H H ⁻:O—H

arrows

product **(1)**

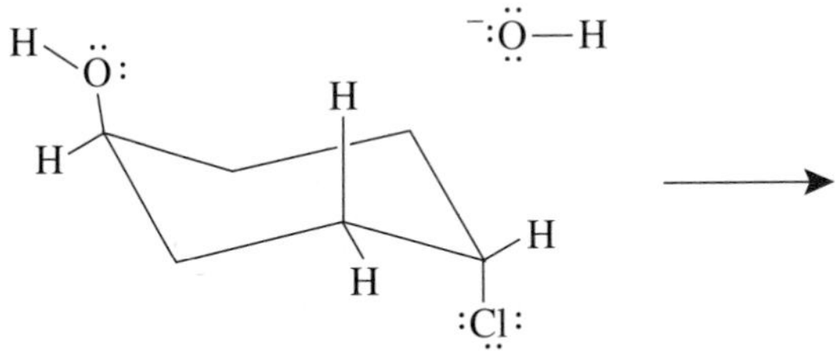

H O: H H ⁻:O—H H H :Cl:

arrows

product **(2)**

What about product (3)? Another ring has formed, so somehow the head must bite the tail (the proper jargon is that this is an intramolecular reaction). Try the trick of studying the relationship between the structures of starting material and product.

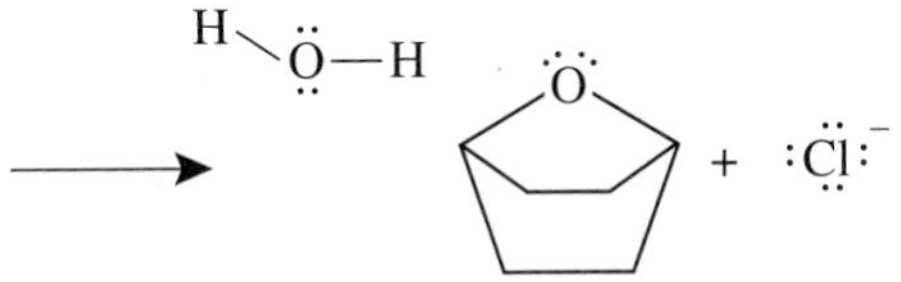

H—O—H + :Cl:⁻

draw starting materials, supply arrows

* * * * *

Here are some problems involving mechanisms traditionally studied later in the course.

9. Using curved arrow notation, show a mechanism for the following.

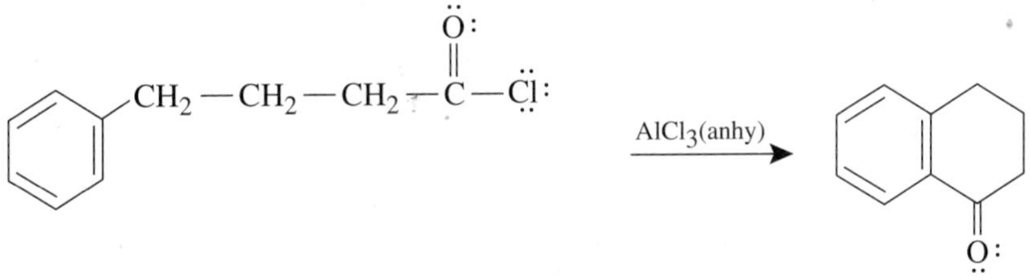

$CH_2—CH_2—CH_2—C(=O)—Cl$ $\xrightarrow{AlCl_3\text{(anhy)}}$

What is going on? You have not lost or gained any carbon atoms. The benzene ring has remained intact. It looks like an intramolecular cyclization. The chemistry is an acid chloride treated with anhydrous aluminum chloride forming a ketone at the aromatic ring. It must be a Friedel-Crafts acylation. Write the mechanism in three steps.

10.

$$\underset{1}{CH_3}-\underset{2}{\overset{\overset{\ddot{O}:}{\|}}{C}}-\underset{3}{CH_2}-\underset{4}{\overset{\overset{\ddot{O}:}{\|}}{C}}-\underset{5}{CH_3} + H_2\ddot{N}-\ddot{N}H_2 \xrightarrow[H_2O]{H^+}$$

H, N:, :N, $CH_3$, $CH_3$

Big picture: the product has five carbons and two nitrogens. Conclude: the two starting molecules have condensed. This is just a twist on problem 7. Circle the remnants of the starting molecules in the product above. Conclude that the two nitrogens have become attached to C2 and C4, respectively. Do you know a reaction in which a nucleophilic amino nitrogen attacks an electrophilic carbonyl carbon? You should. There are other puzzles: the acid catalysis, the ring closure, and the appearance of double bonds in the ring. Try the whole thing.

Here are three mechanism problems where you receive no guidance at all. Try attacking them along the lines suggested in this chapter. My versions of the answers can be found in the answer section. After you have finished, see how closely we agree. You will find plenty of mechanism problems in your text to test your newly acquired skills.

11. This reaction was first reported by Emil Fischer.

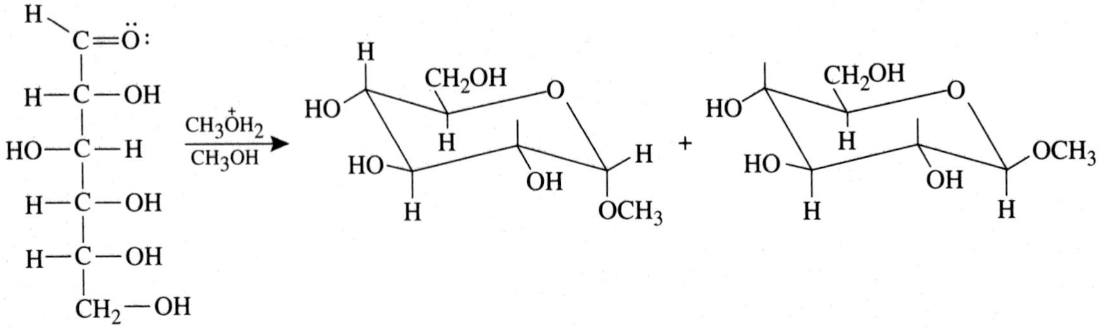

$CH_3\overset{+}{O}H_2$ / $CH_3OH$

H
CH2OH
HO
O
H
HO
OH
H
H
OCH3
+
CH2OH
HO
O
H
HO
OH
OCH3
H
H

12. Your only hint is that a nitrile group is considered a derivative of a carboxylic acid.

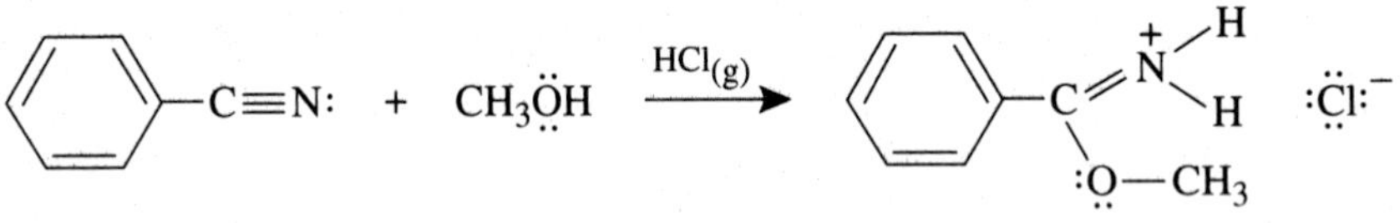

$HCl_{(g)}$

13.

NaOH(aq)

Ph—CH2—C—CH2—Ph + Ph—C—C—Ph

Ph Ph Ph Ph

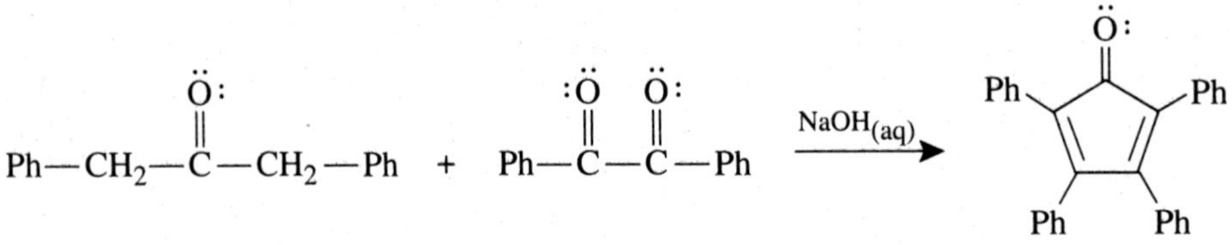

# Answers

## Chapter 4

1.

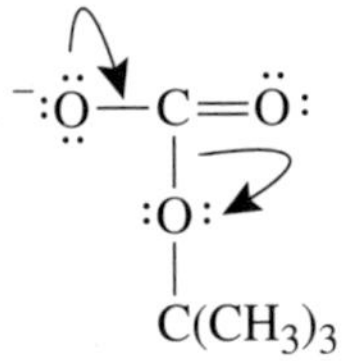

$^{-}:\ddot{O}-C=\ddot{O}:$

$:\ddot{O}:$

$C(CH_3)_3$

2. scribbling, what's going on?, 2, C1, 5, Y

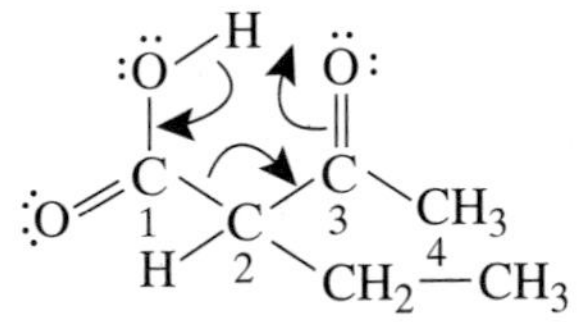

$:\ddot{O}-H$  $\ddot{O}:$

$:\ddot{O}=C$ (1) — C (2) — C (3) — $CH_3$

H (on C2), $CH_2 \overset{4}{-} CH_3$

Are your arrows going clockwise? It gets the job done, but this proton is on a carboxylic acid, and it makes chemical sense that the carboxyl takes the electrons of the O—H bond.

3. 5, oxygen, carbon

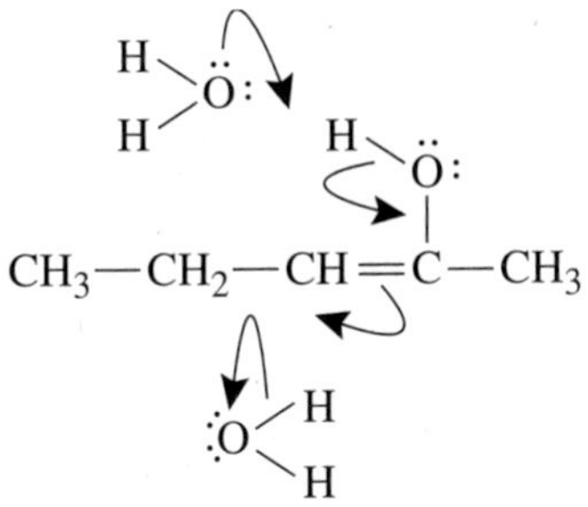

$H_2\ddot{O}:$

$H-\ddot{O}:$

$CH_3-CH_2-CH=C-CH_3$

$:\ddot{O}H_2$

4. (a) $CH_3$ (cyclohexene) + $:\ddot{Br}-\ddot{Br}:$ ⟶ $CH_3$, $+\ddot{Br}:$ + $:\ddot{Br}:^{-}$

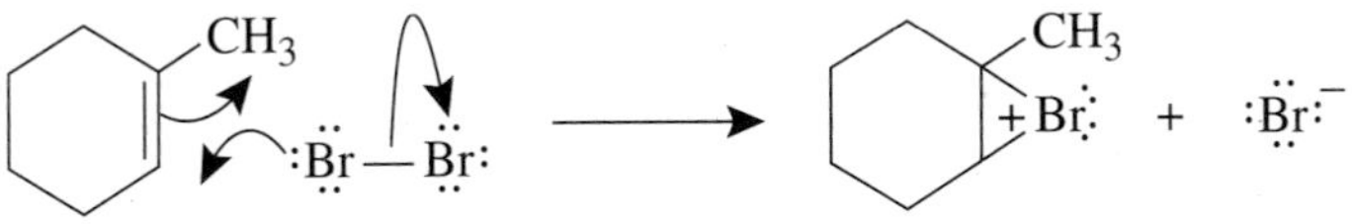

(b) $CH_3$, $+\ddot{Br}:$ + $:\ddot{Cl}:^{-}$ ⟶ $CH_3$, $:\ddot{Cl}:$, $:\ddot{Br}:$

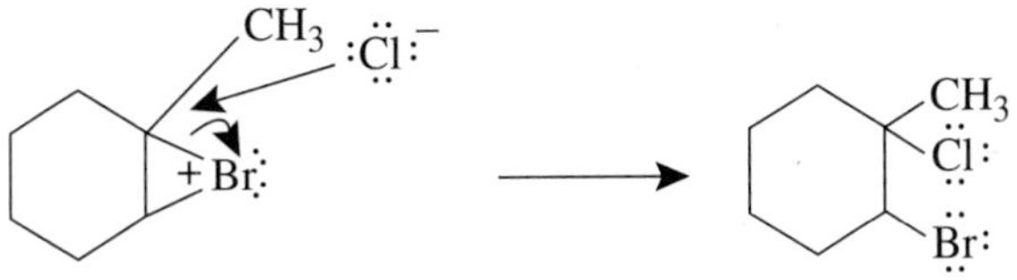

5. C5, C6

(a)

$CH_2{=}CH{-}CH_2{-}CH_2{-}CH_2{-}CH_2{-}\ddot{O}{-}H$ + $H{-}\ddot{O}{-}SO_3H$ → $CH_2(H){-}\overset{+}{C}H{-}CH_2{-}CH_2{-}CH_2{-}\ddot{O}{-}H$ + $^{-}\ddot{O}{-}SO_3H$

(b)

$CH_3{-}\overset{+}{C}H{-}CH_2{-}CH_2{-}CH_2{-}CH_2{-}\ddot{O}{-}H$ → (2-methyltetrahydropyran-O⁺–H, $CH_3$) + $H{-}\ddot{O}{-}H$ → (2-methyltetrahydropyran, $CH_3$) + $H_3O^+$

6. alkene, 5, 3, 4

(a)

$(CH_3)_3C{-}CH_2{-}\ddot{O}{-}H$ + $H{-}\ddot{O}{-}SO_3H$ → $(CH_3)_3C{-}CH_2{-}\overset{+}{O}H_2$ → $(CH_3)_3C{-}\overset{+}{C}H_2$

(b)

$(CH_3)_3C{-}\overset{+}{C}H_2$ → $CH_3{-}\overset{+}{C}(CH_3){-}CH_2{-}CH_3$

(c)

$CH_3{-}\overset{+}{C}(CH_3){-}C(CH_3)(H){-}H$ + $:\ddot{O}^{-}{-}SO_3H$ → product

7. 4, 8,

$CH_3{-}C(CH_3)_2{-}CH{=}C(CH_3)_2$

(a)

$CH_3{-}C(CH_3){=}CH_2$ + $H{-}\ddot{O}{-}SO_3H$ → $CH_3{-}\overset{+}{C}(CH_3){-}CH_3$ + $^{-}\ddot{O}{-}SO_3H$

(b) 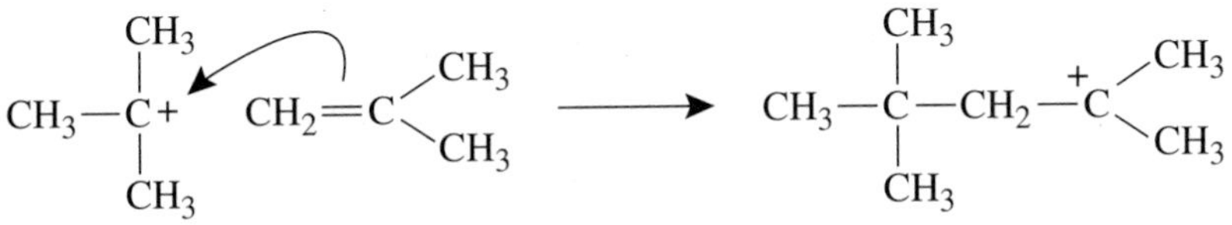 $CH_3-\overset{CH_3}{\underset{CH_3}{C}}{}^+ \quad CH_2{=}C(CH_3)_2 \longrightarrow CH_3-\overset{CH_3}{\underset{CH_3}{C}}-CH_2-\overset{+}{C}(CH_3)_2$

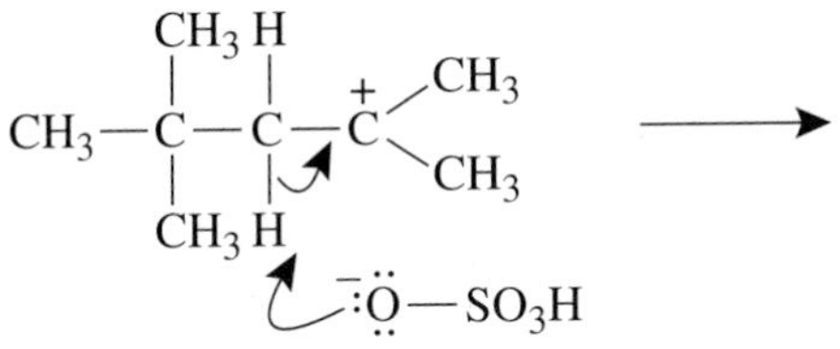

$CH_3-\overset{CH_3}{\underset{CH_3}{C}}-CH_2-\overset{+}{C}(CH_3)_2$ + $^-{:}\ddot{O}-SO_3H$ (removes H from $CH_2$) $\longrightarrow$ major product

$CH_3-\overset{CH_3}{\underset{CH_3}{C}}-CH_2-\overset{+}{C}(CH_3)(CH_2-H)$ + $^-{:}\ddot{O}-SO_3H$ (removes H from $CH_3$) $\longrightarrow$ minor product

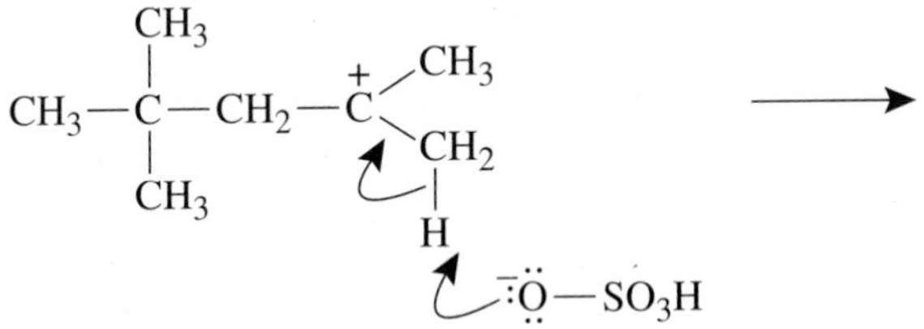

8. inverted, anti

(a) 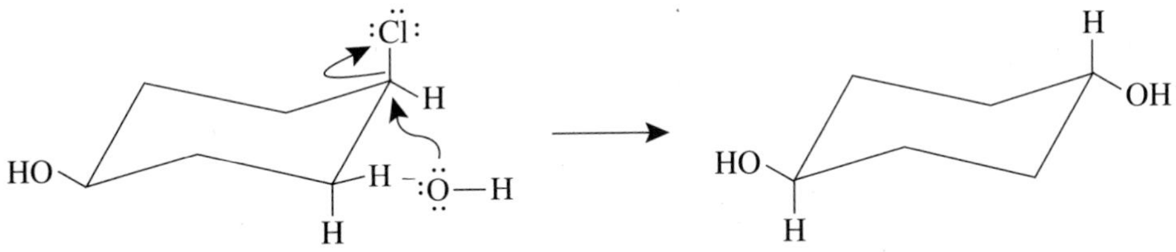 HO–cyclohexyl–Cl + $^-{:}\ddot{O}-H$ $\longrightarrow$ HO–cyclohexyl–OH

(b) 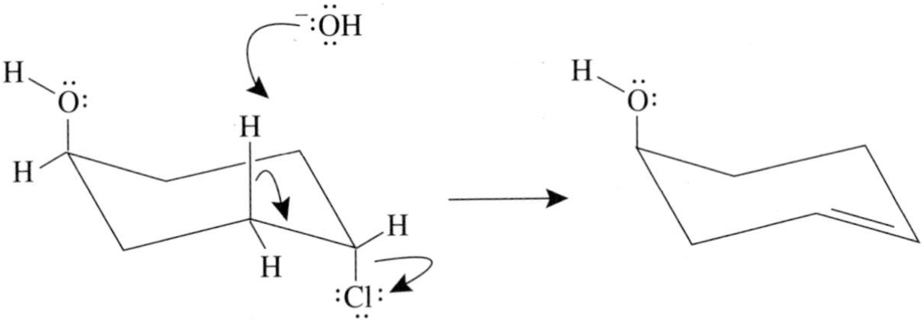 HO–cyclohexyl–Cl + $^-{:}\ddot{O}H$ $\longrightarrow$ HO–cyclohexene

(c) 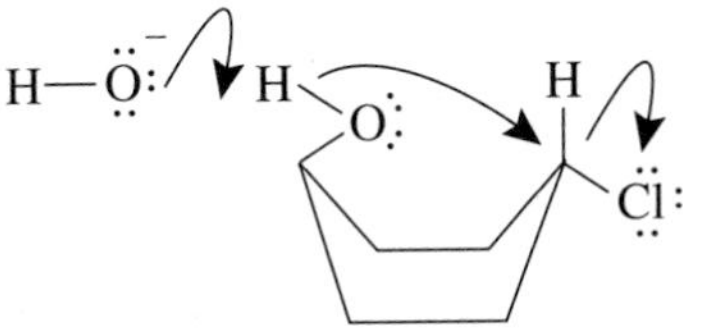 $H-\ddot{O}{:}^-$ + HO–cyclohexyl–Cl

9.

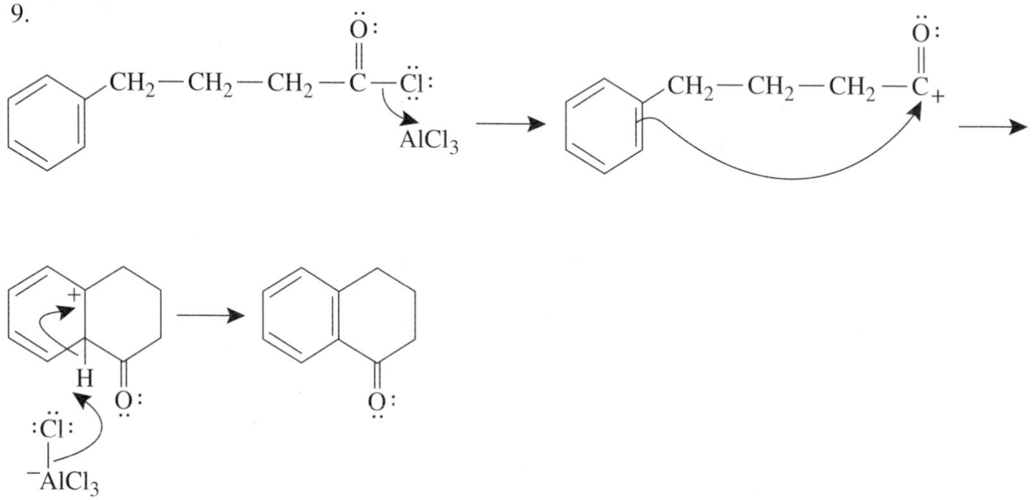

$$\text{C}_6\text{H}_5\text{—CH}_2\text{—CH}_2\text{—CH}_2\text{—C(=O)—Cl} \xrightarrow{AlCl_3} \text{C}_6\text{H}_5\text{—CH}_2\text{—CH}_2\text{—CH}_2\text{—C}^+\text{=O} \longrightarrow$$

H, :Cl:, ⁻AlCl₃, O: →  O:

10. There are *many* steps in this mechanism, but most of them are just proton management. Several parts of this process—the nucleophilic additions to close the ring and the subsequent dehydrations to form the double bonds—are acid-catalyzed. It seems in this answer that protons go on and off oxygen and/or nitrogen with careless ease. It's all right. Proton transfer to and from atoms of these elements has practically no energy barrier. In situations like this, proper proton management is necessary to avoid high-energy species such as two positive charges on the same molecule or a negative charge in an acidic medium.

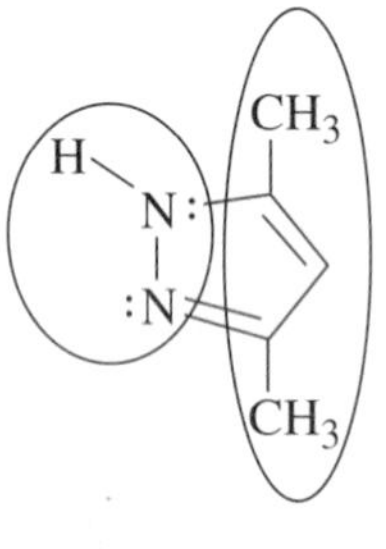

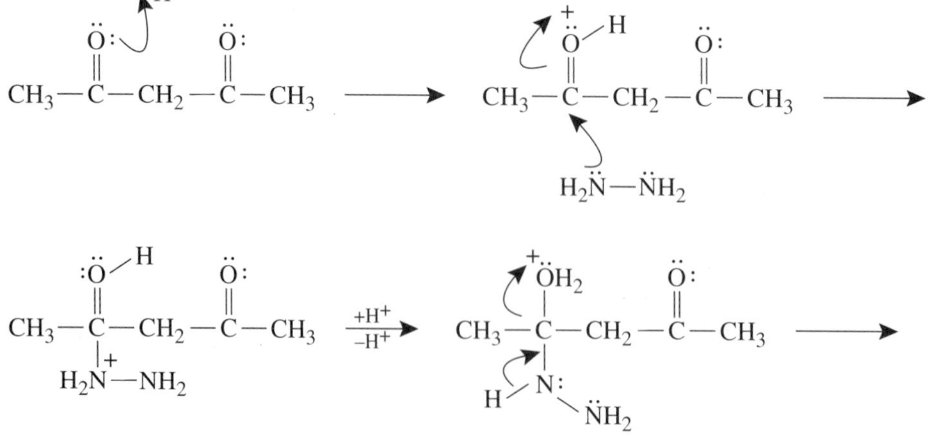

H, N:, :N, $CH_3$, $CH_3$

$$CH_3\text{—C(=O)—}CH_2\text{—C(=O)—}CH_3 \xrightarrow{H^+} CH_3\text{—C(=}\overset{+}{O}\text{H)—}CH_2\text{—C(=O)—}CH_3 \longrightarrow$$

$H_2\ddot{N}\text{—}\ddot{N}H_2$

$$CH_3\text{—C(OH)(}\overset{+}{N}H_2\text{—}NH_2\text{)—}CH_2\text{—C(=O)—}CH_3 \underset{-H^+}{\overset{+H^+}{\longrightarrow}} CH_3\text{—C(}\overset{+}{O}H_2\text{)(NH—}\ddot{N}H_2\text{)—}CH_2\text{—C(=O)—}CH_3 \longrightarrow$$

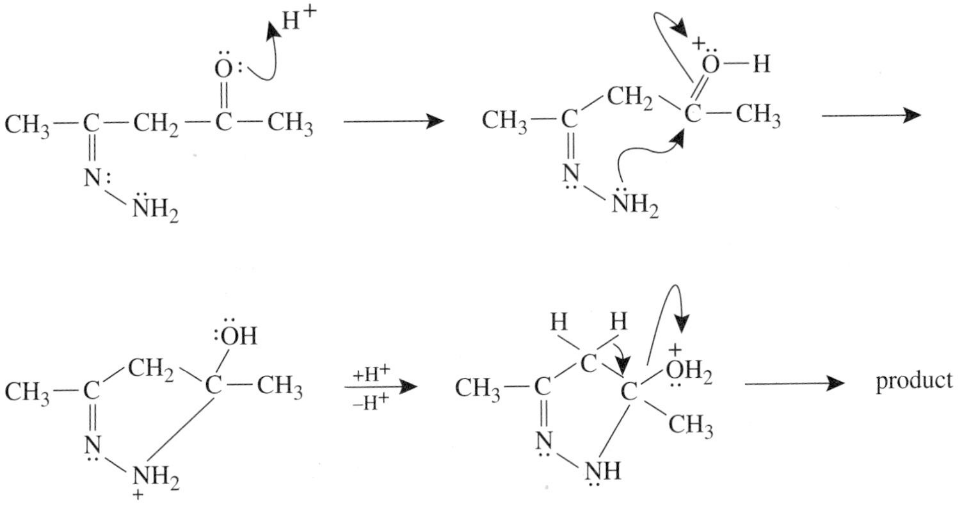

$CH_3$—C(=N—$NH_2$)—$CH_2$—C(=O)—$CH_3$ → $CH_3$—C(=N—$NH_2$)—$CH_2$—C(=$\overset{+}{O}$—H)—$CH_3$ →

$CH_3$—C(=N)—$CH_2$—C(OH)($CH_3$)—$\overset{+}{N}H_2$ $\xrightarrow[-H^+]{+H^+}$ $CH_3$—C(=N)—CH—C($\overset{+}{O}H_2$)($CH_3$)—NH → product

Impatience with the length of this mechanism tempts one to have the two additions occur simultaneously. Doing that is unrealistic for at least two reasons: it would require both carbonyls to be protonated at the same time, and the necessary alignment of the two molecules is entropically unfavorable. There is a question (unanswered, I think) of whether the first dehydration occurs before or after the ring closes. You could make the second double bond a C=N or a C=C. But it has to be C=C to get the right product. There is a reason why it goes that way. It has to do with aromaticity. You try to figure it out.

11. The big picture. It's one of those intramolecular ring closures. Analysis indicates that C1 becomes bound to the oxygen on C5. The $OCH_3$ from the methanol also becomes bound to C1. There are two stereoisomeric products but we are not going to bother with that just yet. The chemistry looks like acid-catalyzed nucleophilic addition to the C=O.

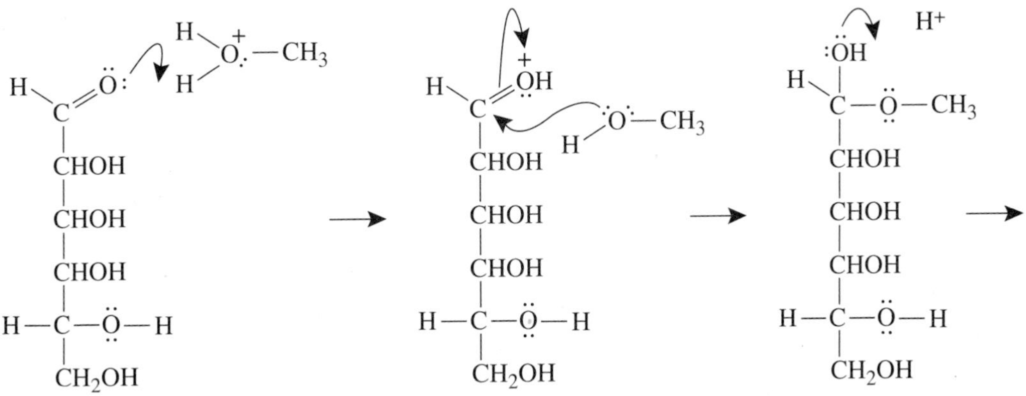

H—C(=O)—CHOH—CHOH—CHOH—H—C—O—H—$CH_2OH$ + $H_2\overset{+}{O}$—$CH_3$ → H—C(=$\overset{+}{O}$H)—CHOH—CHOH—CHOH—H—C—O—H—$CH_2OH$ + H—O—$CH_3$ → H—C(OH)(—O—$CH_3$)—CHOH—CHOH—CHOH—H—C—O—H—$CH_2OH$ + $H^+$ →

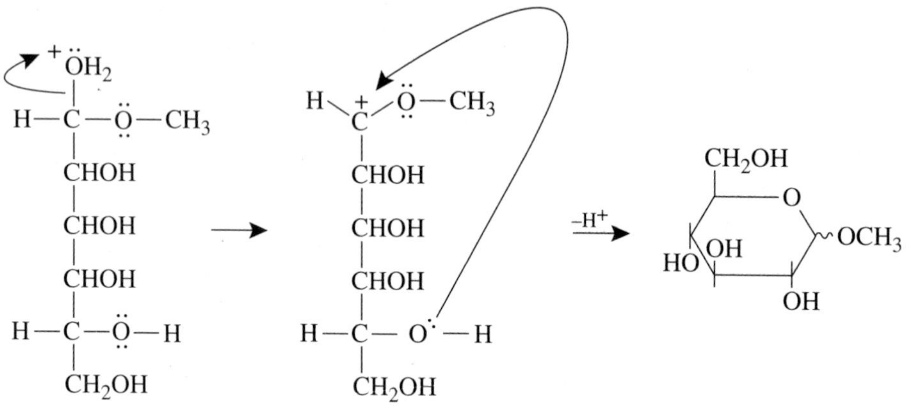

$$H-C(OH_2^+)-\ddot{O}-CH_3 \;(\text{CHOH})_3\; H-C-\ddot{O}-H\; CH_2OH \longrightarrow H-\overset{+}{C}-\ddot{O}-CH_3\;(\text{CHOH})_3\; H-C-\ddot{O}-H\; CH_2OH \xrightarrow{-H^+}$$

stereochemistry?

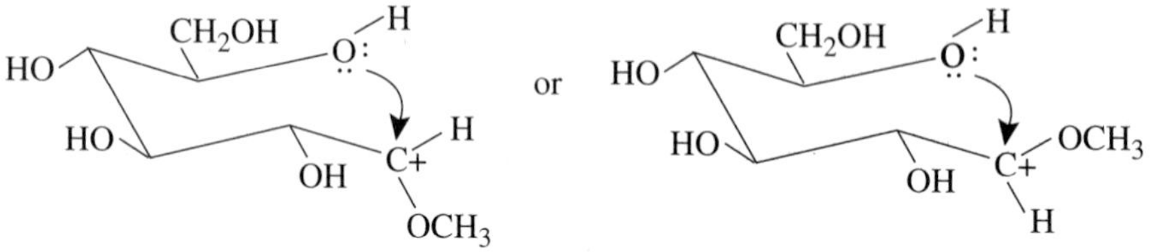

or

12. The hint should be all you needed. This is just an unusual nucleophilic acyl substitution.

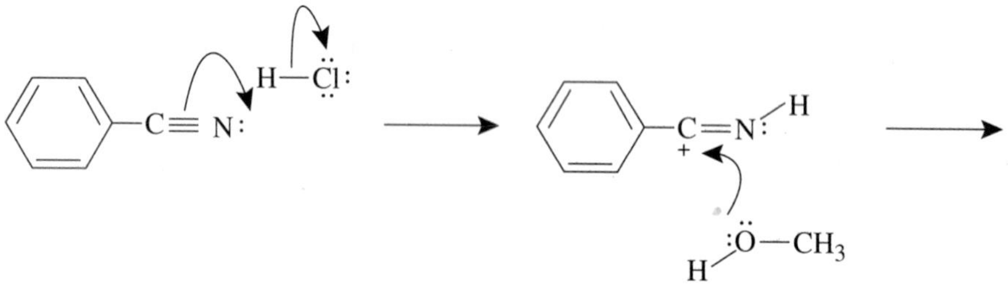

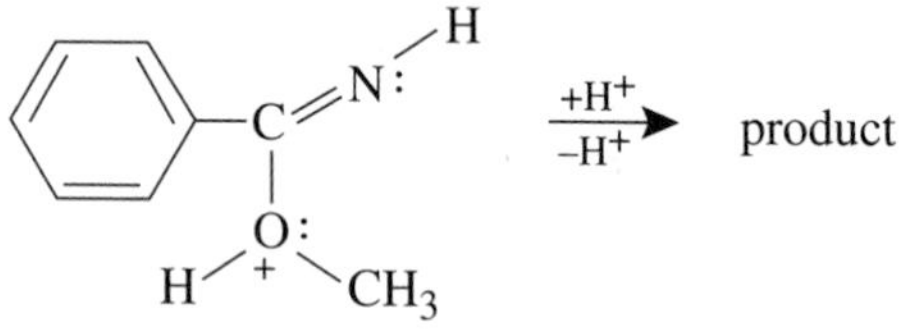

$$\xrightarrow{+H^+}_{-H^+} \text{product}$$

13. This ring is formed by the condensation of two molecules reminiscent of problem 10. By encircling the starting materials in the structure of the product, one concludes that the —$CH_2$— groups

in one have become bound to the carbonyl carbons of the other. NaOH strongly suggests aldol-type chemistry.

$H\ddot{O}^{-}$: + $PhCH_2C(=O)CH_2Ph$ → $PhCH_2C(=O)\ddot{C}^{-}HPh$ + $PhC(=O)C(=O)Ph$ $\xrightarrow{+H^+}$ $PhCH_2C(=O)CH(Ph)C(OH)(Ph)C(=O)Ph$ $\xrightarrow[\text{spontaneous}]{-H_2O}$

$PhCH_2C(=O)C(Ph)=C(Ph)C(=O)Ph$ $\xrightarrow{\text{same steps}}$ product

# 5 Some Reactions from Biochemistry

In many academic programs, the chemistry course that follows organic is biochemistry. Indeed, most modern texts for the first course in organic chemistry conclude with several chapters on the chemistry of biochemically important classes of compounds such as proteins, carbohydrates, lipids, and nucleic acids.

It is not an entirely foolish conceit to say that biochemistry is organic chemistry occurring in more complex systems. Certainly, a thorough knowledge of organic is crucial for those who would know and use biochemistry.

In this section I shall introduce examples from bioorganic chemistry and elementary biochemistry. I hope to make two points: biochemical reaction mechanisms are not fundamentally different from organic mechanisms, and the skills that you learned in earlier sections of this book can be applied in biochemistry.

## Bioorganic Reactions

We begin with a selection of reactions that are not strictly biochemical; rather they are organic reactions used in the service of studying biochemistry. You must be pretty far along in the course by now, so I feel safe in initiating some shorthand notation that you have no doubt seen in your text or in lecture. Also, I am omitting some unshared pairs of electrons from atoms that are remote to the site of the action. Where arrows are supplied, draw the products. Where the products are given, supply arrows. (In this chapter the molecules are larger and more complex; in some cases, you may have to squeeze your answers into the space provided.)

1. A step in the laboratory synthesis of a steroid

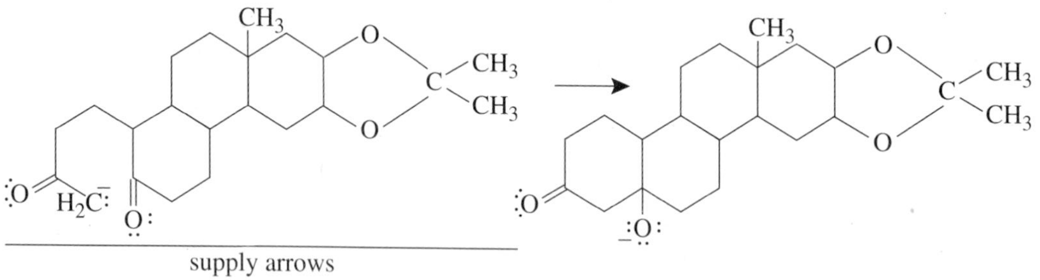

$CH_3$ O C $CH_3$ $CH_3$ O :O $H_2C$: O: → $CH_3$ O C $CH_3$ $CH_3$ O :O :O:

supply arrows

2. A step in the laboratory synthesis of an amino acid

$$CH_3-\overset{\overset{:\ddot{O}}{\|}}{C}-\ddot{N}H-\underset{COOCH_2CH_3}{\overset{COOCH_3CH_3}{C:^-}} + CH_2(C_6H_5)-\ddot{Br}: \longrightarrow \underline{\qquad\qquad} + :\ddot{Br}:^-$$

3a. Two consecutive steps that protect an amino acid for peptide synthesis

$$(CH_3)_3C-\ddot{O}-\overset{\overset{:\ddot{O}}{\|}}{C}-\ddot{O}-\overset{\overset{:\ddot{O}}{\|}}{C}-\ddot{O}-C(CH_3)_3 + H_2\ddot{N}-CH(CH_3)-CO_2^- \longrightarrow (CH_3)_3C-\ddot{O}-\underset{H_2N^+-CH(CH_3)-CO_2^-}{\overset{:\ddot{O}:^-}{C}}-\ddot{O}-\overset{\overset{:\ddot{O}}{\|}}{C}-\ddot{O}-C(CH_3)_3$$

supply arrows

3b.

$$(CH_3)_3-C-\ddot{O}-\underset{\underset{CH_3-CH-COO^-}{:NH}}{\overset{:\ddot{O}:^-}{C}}-\ddot{O}-\overset{\overset{:\ddot{O}}{\|}}{C}-\ddot{O}-C(CH_3)_3 \longrightarrow \underline{\qquad} + \underline{\qquad}$$

4. A step in the laboratory synthesis of a peptide

$$(CH_3)_3C-O-\overset{\overset{O}{\|}}{C}-NH-CH_2-\overset{\overset{:\ddot{O}}{\|}}{C}-\ddot{O}:^- + C_6H_{11}-\ddot{N}=C=\ddot{N}-C_6H_{11} \longrightarrow \underline{\qquad\qquad}$$

5. A step in the synthesis of a dinucleotide

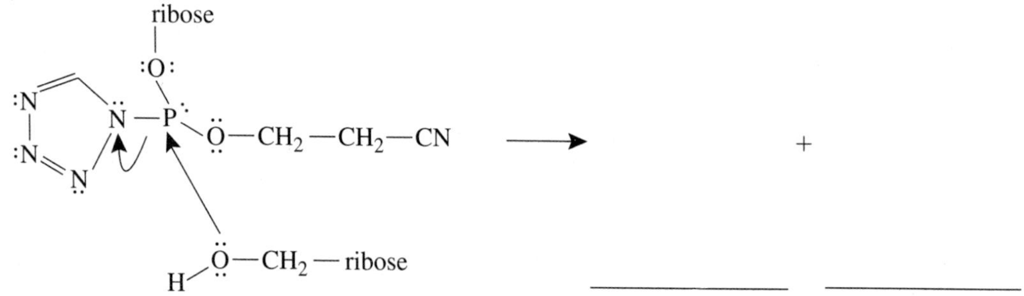


6. A deprotection step subsequent to 5

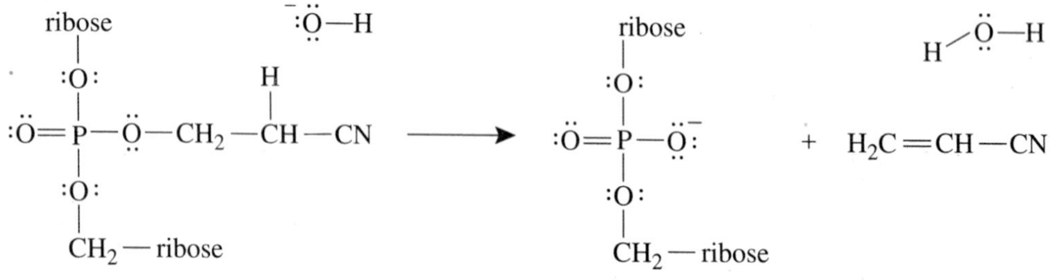


supply arrows

7. Three consecutive steps in the preparation of a sugar derivative

(a)

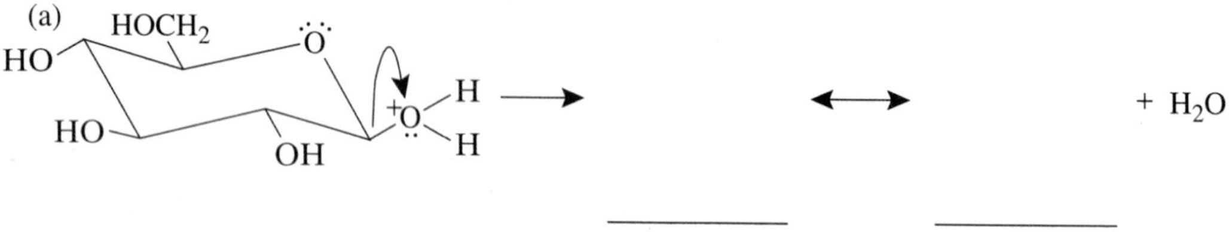


(b)

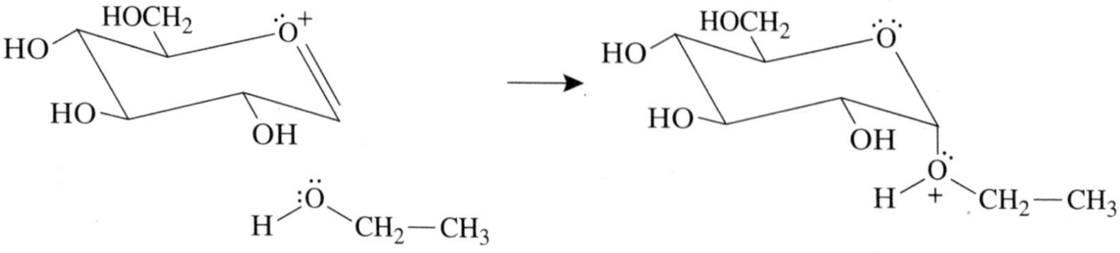


supply arrows

(c)

HOCH$_2$ HO HO OH H + $CH_2-CH_3$ H O: $CH_2-CH_3$

→ ______ + ______

8. A step in the synthesis of a disaccharide

R O O O O $CH_2$ O: H :O: O O $CH_2$ O: O

→ ______

And now some reactions that occur—with the help of enzymes (see below)—in living cells. The first four reactions are from glycolysis, the metabolism of blood sugar.

9.

$CH_2-OPO_3^=$ | C=O: | HO—CH | HC—OH | HC—OH | $CH_2-OPO_3^=$ → $CH_2-OPO_3^=$ | C=O: | HO—C—H HC=$\overset{+}{O}$H | HC—OH | $CH_2-OPO_3^=$

______
supply arrows

The element phosphorus is a big player in biochemistry. In the exercises that follow, phosphorus is in the phosphate state of oxidation. Treat it as you would a carboxylate carbon. Sometimes phosphorus ends up with ten valence electrons (see page 12).

10.

adenine —O—P—O—P—O—P ⟶ adenine —O—P—O—P—O

CH$_2$ HO HO OH OH O

+ ______

11. (a)

O—PO$_3^=$ H :B B: H—O O—PO$_3^=$

B$^+$—H CHOH H ⟶ C O—H H—B$^+$

CH$_2$—O—PO$_3^=$ CHOH

CH$_2$—O—PO$_3^=$

______

supply arrows

(b)

O H C O—P—O OH :O BH$^+$

B: CHOH ⟶ + O—P—O

CH$_2$—O—PO$_3^=$ ______ :O:

And now three from the urea cycle

12.

H$_2$N: C—NH$_2$

(CH$_2$)$_3$ O ⟶

H$_3\overset{+}{N}$—CH PO$_3^=$ ______

COO$^-$

13.

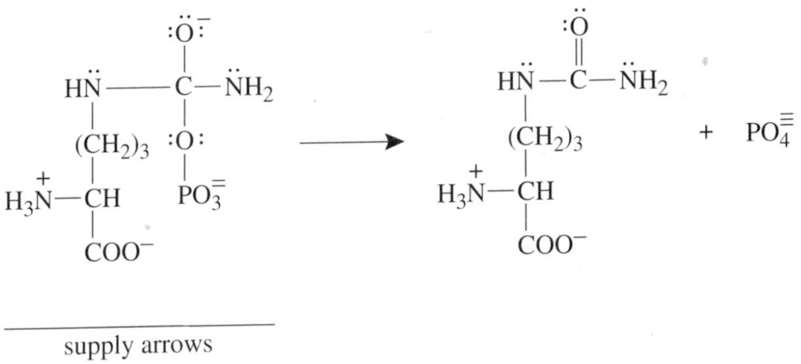

$$:\ddot{O}:^{-}$$
$$H\ddot{N}—C—\ddot{N}H_2$$
$$(CH_2)_3 \quad :\ddot{O}:$$
$$H_3\overset{+}{N}—CH \quad PO_3^{=}$$
$$COO^-$$

$$\longrightarrow$$

$$:\ddot{O}$$
$$H\ddot{N}—C—\ddot{N}H_2$$
$$(CH_2)_3 \qquad + \quad PO_4^{\equiv}$$
$$H_3\overset{+}{N}—CH$$
$$COO^-$$

______________________
supply arrows

14.

$$\overset{+}{B}—H$$
$$H \diagdown \overset{+}{N} \diagup H \qquad H \qquad COO^-$$
$$H\ddot{N}—C—\ddot{N}H—CH$$
$$(CH_2)_3 \qquad H—CH$$
$$H_3\overset{+}{N}—CH \qquad \ddot{B} \qquad COO^-$$
$$COO^-$$

$$\longrightarrow$$

$$\ddot{B}$$

$$\overset{+}{B}—H$$

______________ ______________

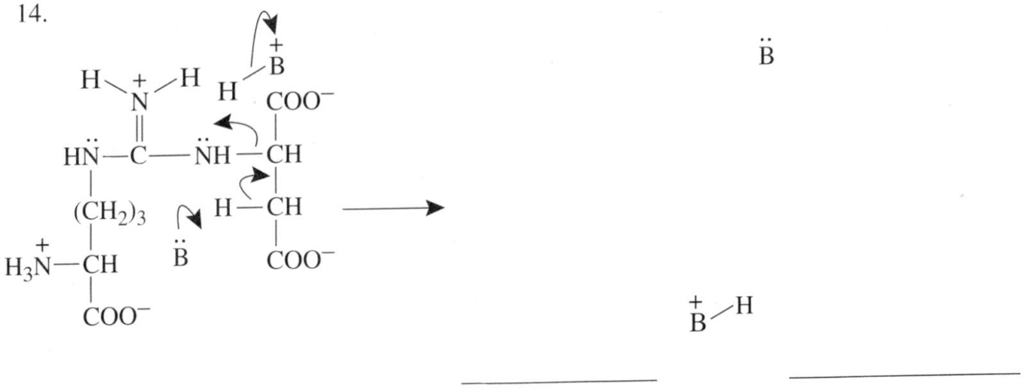

Please notice that, although the molecules are larger and more complex, you have used the same symbolism you have been using all along to complete these exercises correctly.

## Enzymes

Reactions that occur in living cells are almost always catalyzed by enzymes. To chemists of the author's age, i.e. pre-war, enzymes are marvelous. Thirty years ago, I heard a friend of mine describe biochemistry as follows, "First, the enzyme and the substrate combine to form an enzyme-substrate complex, and then a miracle occurs." Since then, much has been learned about how these wonderful natural catalysts work. Our knowledge of organic mechanisms has been a major player in these discoveries.

Enzymes are efficient and, in most cases, remarkably specific. Being proteins, enzymes are large, complex molecules that are produced within the cell (*via* reactions that are catalyzed by *other* enzymes). Like all catalysts, they lower the activation energy of the reaction. They do so by using many tricks such as lowering entropy by bringing reactants together in perfect alignment, inducing steric strain in a reactant, providing an acid and/or base at just the right place, providing an oxidant and/or reductant at just the right place, using an easier pathway from reactant to product, and so on.

The examples and problems that follow are from enzyme-catalyzed reactions. I use symbolism that abbreviates for the enzyme and shows only the reactive sites, a common practice in biochemistry.

**Chymotrypsin** is a protease, i.e., an enzyme that catalyzes the hydrolysis of proteins. There are many proteases. This one is a digestive enzyme. Study this lengthy illustration to prepare for the exercises that follow.

The task of chymotrypsin is to cleave a protein into smaller fragments by catalyzing the hydrolysis of some of the peptide (amide) bonds in the chain of amino acids that constitute the protein. As with all catalysts, the enzyme itself emerges unchanged from the process. Nevertheless, it is very busy *during* the process.

peptide bond

$$\underbrace{R-\overset{\overset{\displaystyle :\ddot{O}}{\|}}{C}-\ddot{N}H-R'}_{\substack{\text{long protein chain}\\ \text{(with many peptide bonds)}}} \xrightarrow[\text{chymotrypsin}]{H_2O} \underbrace{R-\overset{\overset{\displaystyle :\ddot{O}}{\|}}{C}-\ddot{O}-H \quad + \quad H_2N-R'}_{\text{two smaller polypeptide fragments}}$$

The astonishing efficiency occurs because the enzyme provides proton donors and proton acceptors in just the right places at just the right times, thus avoiding the appearance of naked + or − charge. The acid and base centers are functional groups from the side chains of certain of the amino acids in the protein.*

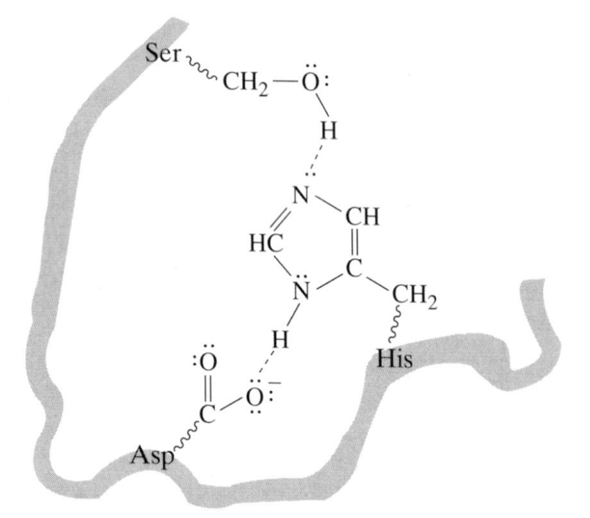


This is an oversimplified drawing of the active site of the enzyme. The shaded area represents the bulk of the enzyme. *Ser* stands for the amino acid, serine, the side chain of which contributes a hydroxymethyl group. *Asp* is for aspartic acid, which contributes the carboxylate anion. *His* is for histidine, which contributes its imidazole side chain. In the first drawing, dotted lines for the hydrogen bonds are shown. The hydrogen bonds are crucial in holding the active site together, but to avoid unnecessary clutter they will not be shown in the illustrations.

---

*There is a wonderful opportunity for creating confusion here. Both the substrate (the enzyme's target) and the catalyst (the enzyme itself) are proteins!

Amide (peptide) bonds are hard to hydrolyze. Chymotrypsin overcomes the difficulty by first converting the amide to an ester and then hydrolyzing the ester, a much easier task.

An illustration: Converting the amide to an ester.

The enzyme uses the serine —OH to bind the substrate. As the nucleophilic attack occurs, the —OH proton is transferred, making the serine oxygen more nucleophilic without ever having to gain a full negative charge.

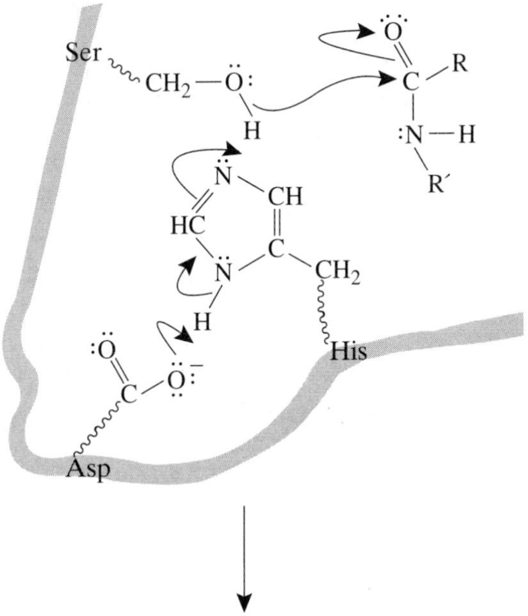

Ser $CH_2$—O: H :N CH HC C N $CH_2$ H His :O C O: Asp O C R :N—H R′

Now the ester forms, bound to the enzyme, and most importantly, the peptide C—N bond breaks. Once again, the simultaneous transfer of a proton eliminates the appearance of a full negative charge on the nitrogen.

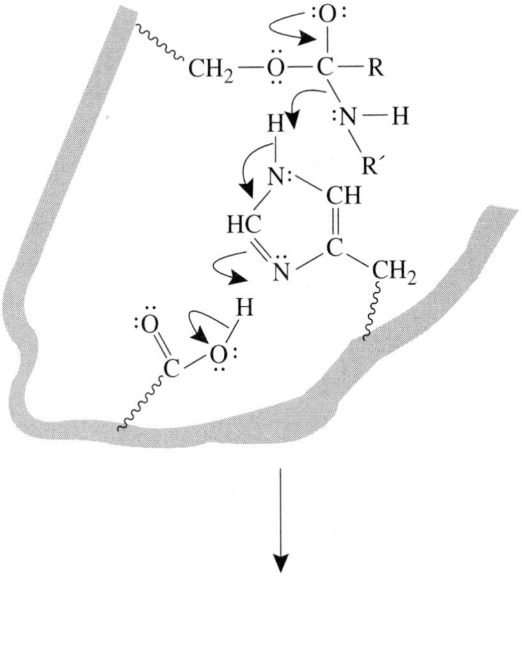

:O: $CH_2$—O—C—R H :N—H R′ N: CH HC C N $CH_2$ H :O C O:

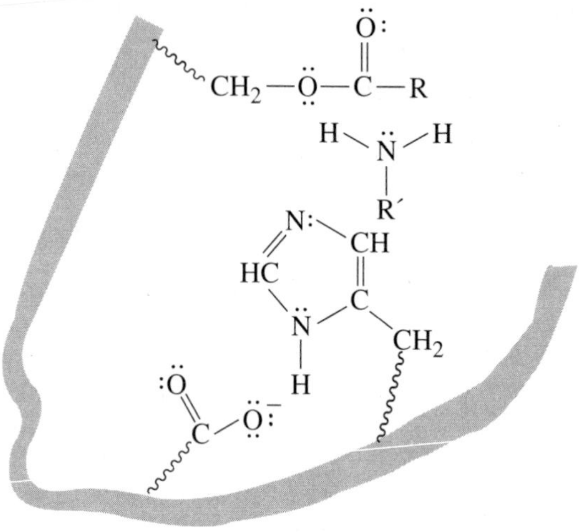

The N-terminal polypeptide fragment is now fully formed and is released for further digestion and metabolism.

15. An exercise: Using arrows, show how the serine ester is converted to the C-terminal polypeptide fragment and the restored enzyme.

(a)

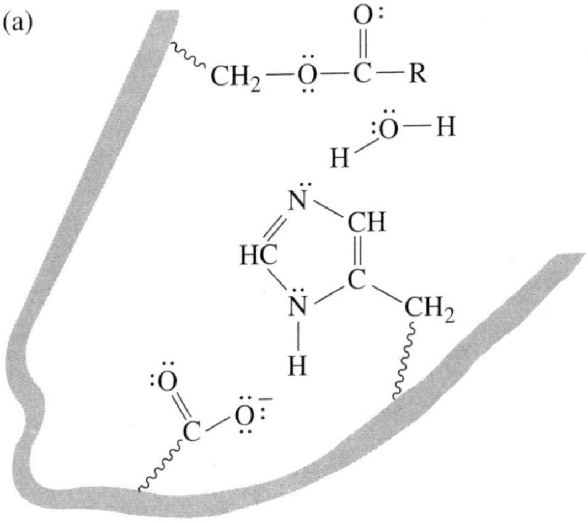

A water molecule has taken the place of the N-terminal fragment (enzymes can do that). In almost the exact reverse of the illustration above, push electrons to complete the enzymatic process.

supply arrows

(b)

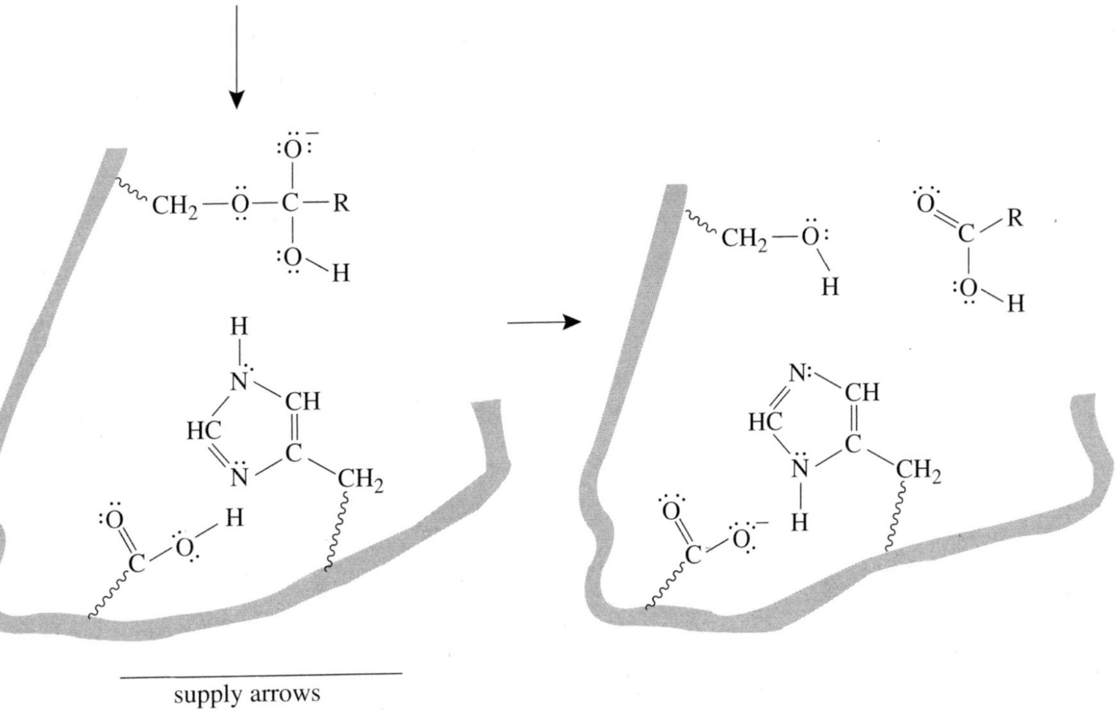

supply arrows

Exercise 9 (page 185) is a useful example of learning to push electrons. However, it is an egregious oversimplification of a metabolic reaction in which a six-carbon sugar is cleaved into two three-carbon fragments. That product in exercise 9, the free enolate anion, just would not form in a cell nor would the other product, the protonated aldehyde. This important reaction is catalyzed by the enzyme **aldolase** in a way that avoids both these high-energy intermediates. Aldolase uses the side chain of a lysine (lys) to bind the substrate.

16. (a)

Lys, $\overset{+}{H}N{=}C$, $CH_2{-}OPO_3^{=}$, HO—CH, HC—Ö—H, CHOH, $CH_2{-}OPO_3^{=}$, Cys, His, $CH_2$

We join the action after the enzyme-substrate complex has formed. A cysteine anion and a histidine cation stand ready to act as base and acid, respectively.

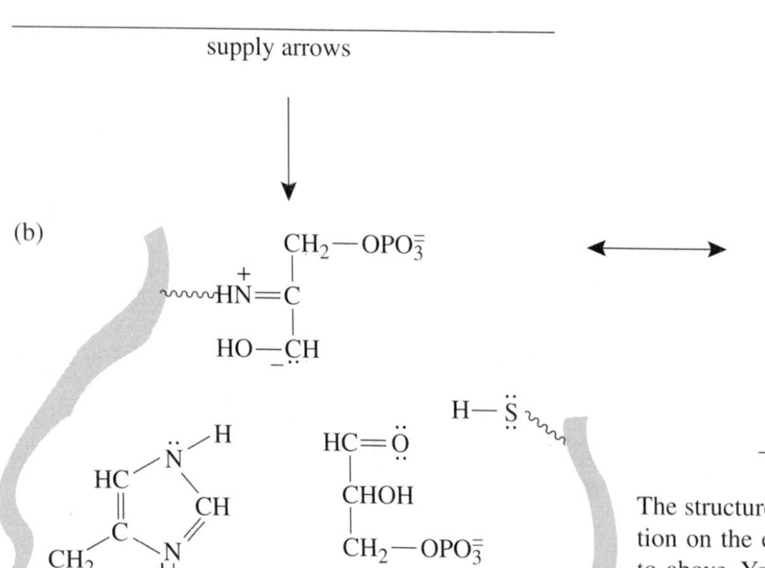


resonance structure

The structure to the left is a variation on the enolate anion referred to above. You ought to be able to draw a resonance structure that shows how the enzyme stabilizes it.

17.

In the next step, a proton transfer neutralizes the anion.

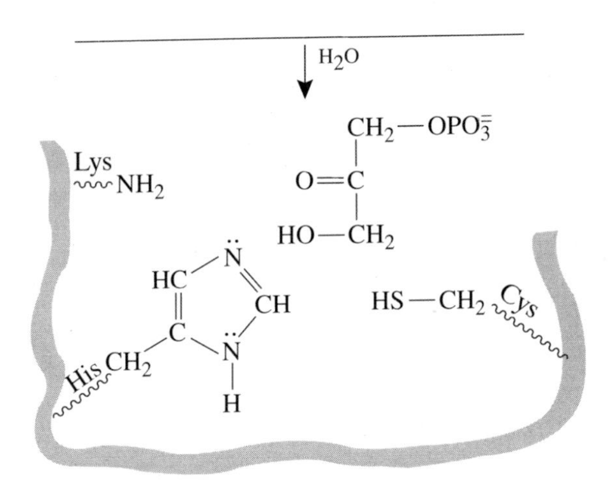


Hydrolysis yields the product and returns the enzyme to its original state.

## Coenzymes

In the two previous examples of catalysis by an enzyme, we have seen the enzyme bind to the substrate in some productive way and then provide acid-base catalysis. Now we encounter an enzyme that uses a coenzyme. A coenzyme is a small molecule, originating from a vitamin, that is associated with, but not part of, the enzyme. The coenzyme is altered during the reaction, but the cell finds a way to return it to its original state so that it can be used again.

One of the products of the aldolase-catalyzed reaction is glyceraldehyde-3-phosphate. The cell next oxidizes (dehydrogenates) the aldehyde to the carboxylic acid state of oxidation and releases it as the phosphate ester, 1,3-diphosphoglycerate. The enzyme, **glyceraldehyde-3-**

**phosphate dehydrogenase,** uses an —SH group from the side chain of cysteine, and it uses the coenzyme, nicotinamide adenine dinucleotide, ($NAD^+$, which originates as niacin) as an oxidizing agent.

18.

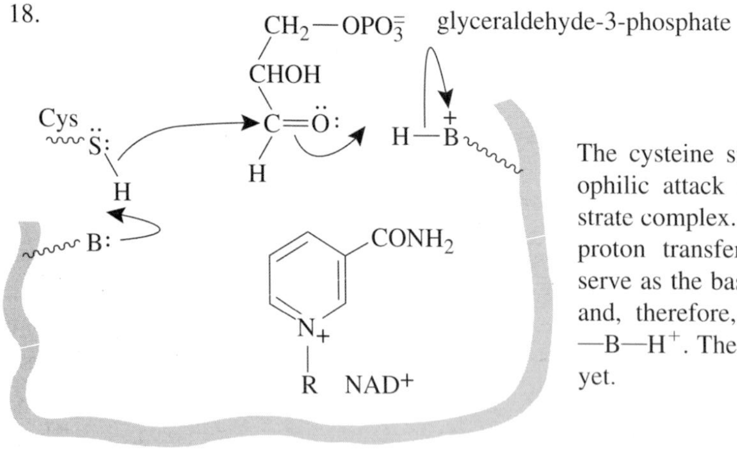

$CH_2$—$OPO_3^=$ glyceraldehyde-3-phosphate

CHOH

Cys

C=Ö:

H—$\overset{+}{B}$

H

B:

$CONH_2$

$N_+$

R $NAD^+$

The cysteine side chain makes a nucleophilic attack to form the enzyme-substrate complex. The process is assisted by proton transfers. The side chains that serve as the base and acid are not known and, therefore, are written as —B and —B—$H^+$. The coenzyme is not involved yet.

↓

---

19.

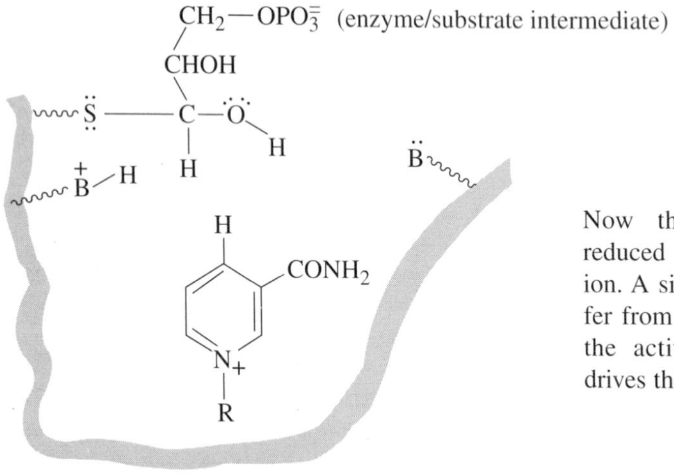

$CH_2$—$OPO_3^=$ (enzyme/substrate intermediate)

CHOH

S —— C—O

H

$\overset{+}{B}$—H H

B̈

H

$CONH_2$

$N_+$

R

Now the coenzyme becomes reduced by accepting an hydride ion. A simultaneous proton transfer from the alcohol to a base on the active site of the enzyme drives the process.

---

supply arrows

↓

$$\text{enzyme–}\ddot{S}\text{–C(=}\ddot{O}\text{:)–CHOH–CH}_2\text{–OPO}_3^{=} \quad (\overset{+}{B}\text{–H},\ \text{NADH},\ \text{H–}\overset{+}{B}) \xrightarrow{HOPO_3^{=}} \;^{=}O_3P\text{–O–C(=}\ddot{O}\text{:)–CHOH–CH}_2\text{–OPO}_3^{=} + \text{enzyme-SH}$$

1, 3-diphosphoglycerate

Note that the erstwhile aldehyde carbon has become a thiol ester carbon, i.e., it has been oxidized.

20. In Exercise 19, show a way that the enzyme might promote the release of the product, 1,3-diphosphoglycerate. You may assume that $HOPO_3^{=}$ is available and that side chain acid and base centers are present. (Use chymotrypsin as a model, and use your own paper.)

* * * * *

Pyridoxal phosphate, from vitamin $B_6$, serves as a coenzyme in a number of different reactions that manage amino acid metabolism. The **transaminases** use pyridoxal phosphate to catalyze reactions in which the $-NH_2$ group from an amino acid is transferred to α-ketoglutarate. The amino acid is converted to a keto acid which is metabolized, and α-ketoglutarate is converted to glutamate and sent into the urea cycle.

$$\underset{\text{amino acid}}{R\text{—}CH(^{+}NH_3)\text{—}COO^{-}} + \underset{\alpha\text{–ketoglutarate}}{^{-}OOC\text{—}CH_2\text{—}CH_2\text{—}C(=O)\text{—}COO^{-}}$$

$$\xrightarrow{\text{enz}} \underset{\alpha\text{–ketoacid}}{R\text{—}C(=O)\text{—}COO^{-}} + \underset{\text{glutamate}}{^{-}OOC\text{—}CH_2\text{—}CH_2\text{—}CH(^{+}NH_3)\text{—}COO^{-}}$$

21.

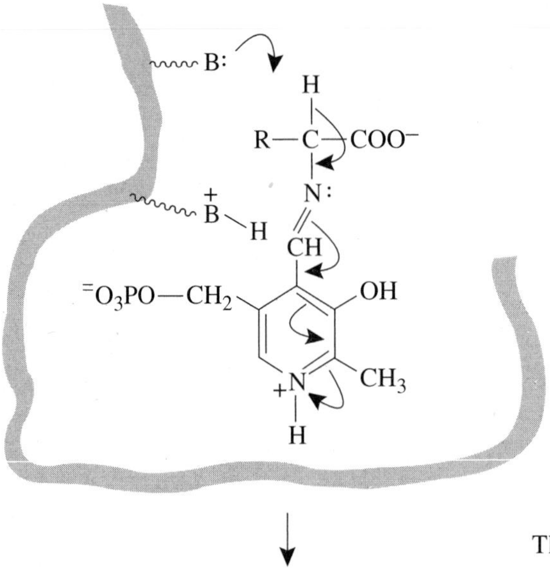

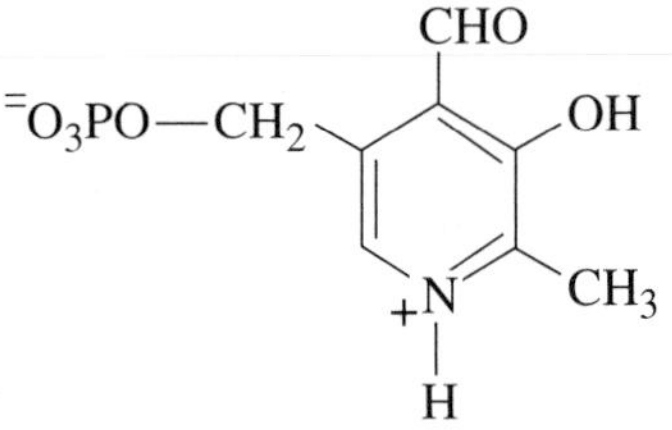

pyridoxal phosphate

We join the action after the enzyme-substrate complex is formed. This time the coenzyme is the culprit in binding the substrate.

The α-carbon of the erstwhile amino acid has been oxidized.

22.

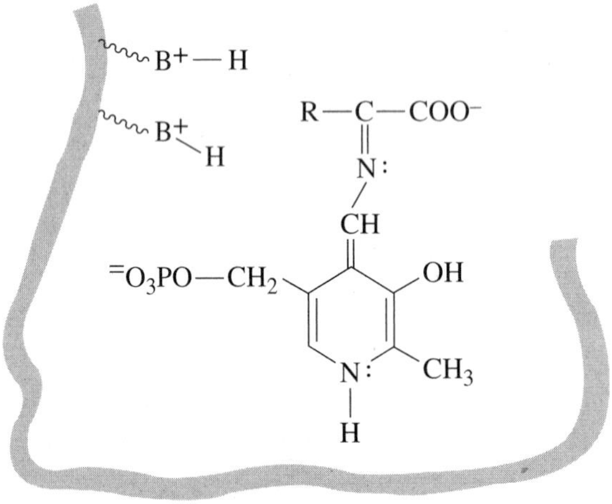

supply arrows

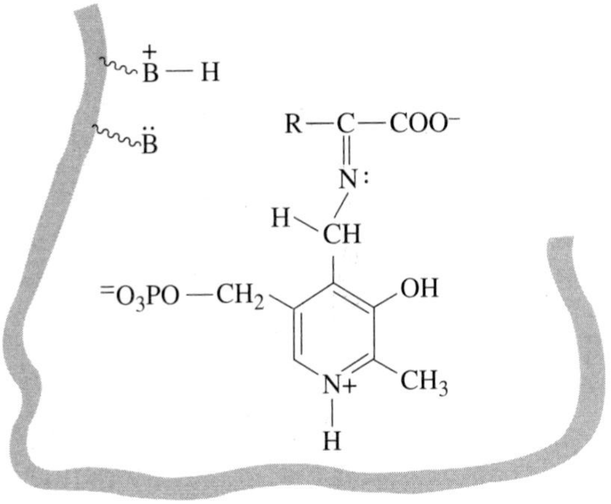

$$\sim\overset{+}{B}-H \qquad \sim\ddot{B} \qquad R-\underset{\underset{\ddot{N}\text{:}-CH(H)-\text{pyridinium}}{\|}}{C}-COO^-$$

$^{=}O_3PO-CH_2$–(pyridinium ring with OH, $CH_3$, N+–H)

The erstwhile aldehyde carbon of pyridoxal phosphate has been reduced.

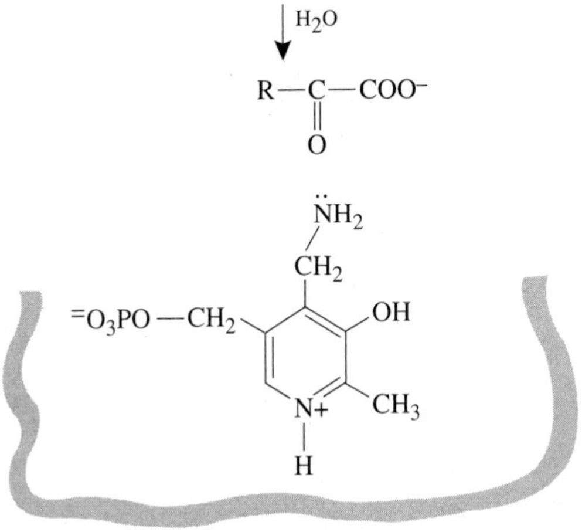

$$\xrightarrow{H_2O}$$

$$R-\underset{\underset{O}{\|}}{C}-COO^-$$

$\ddot{N}H_2$–$CH_2$–(pyridinium ring with $^{=}O_3PO-CH_2$, OH, $CH_3$, N+–H)

As this point, the enzyme expels the α-ketoacid and replaces it with α-ketoglutaric acid. The coenzyme is in its reduced form, pyridoxamine phosphate.

23. Propose a way in which the enzyme can complete the process. That is, from the enzyme-substrate complex below, provide a mechanism to produce glutamate and restore the coenzyme to its original, oxidized state.

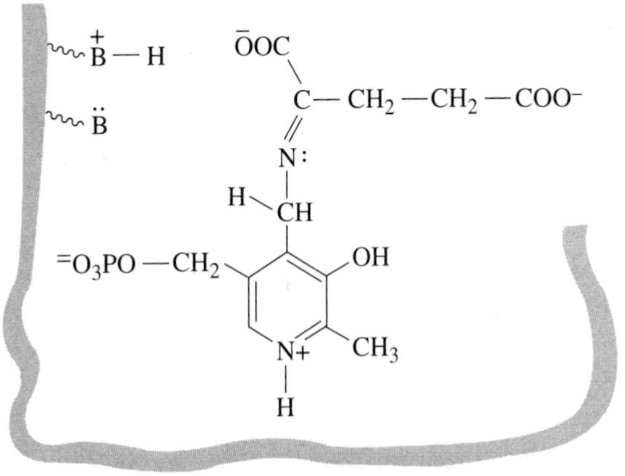

$$\sim\overset{+}{B}-H \qquad \sim\ddot{B} \qquad {}^-OOC-C(=\ddot{N}\text{:}-CH(H)-\text{pyridinium})-CH_2-CH_2-COO^-$$

$^{=}O_3PO-CH_2$–(pyridinium ring with OH, $CH_3$, N+–H)

# Answers

## Chapter 5

1.

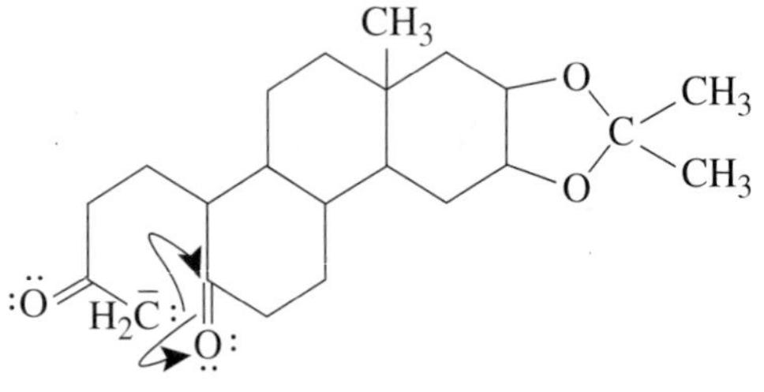

$CH_3$   O   $CH_3$   C   $CH_3$   O

:O   $H_2\bar{C}$:   O:

2.

$$CH_3-\overset{\overset{:\ddot{O}}{\|}}{C}-\ddot{N}H-\underset{COOCH_2CH_3}{\overset{COOCH_2CH_3}{C}}-CH_2-C_6H_5$$

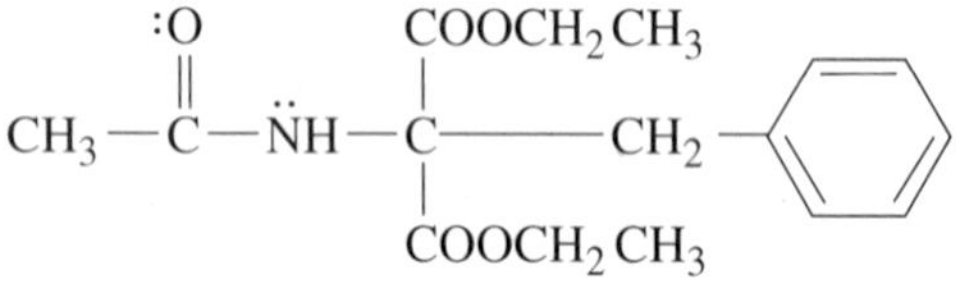

3. (a)

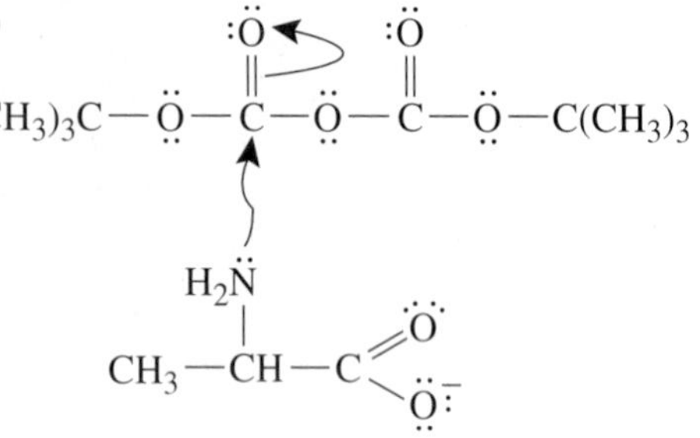

$$(CH_3)_3C-\ddot{O}-\overset{\overset{:\ddot{O}}{\|}}{C}-\ddot{O}-\overset{\overset{:\ddot{O}}{\|}}{C}-\ddot{O}-C(CH_3)_3$$

$$CH_3-\overset{H_2\ddot{N}}{CH}-C\begin{matrix}\nearrow \ddot{O} \\ \searrow \ddot{O}:^-\end{matrix}$$

(b)

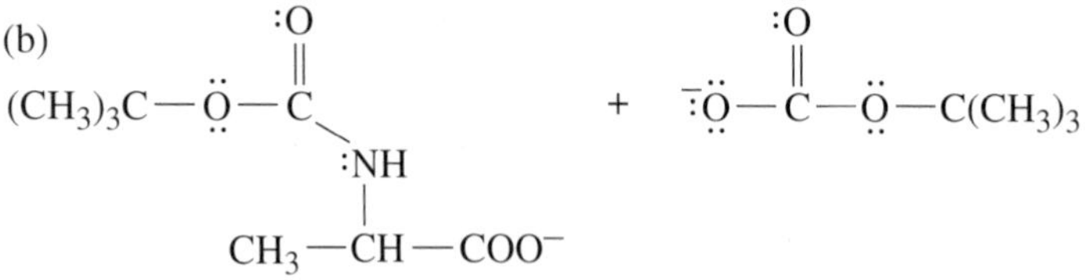

$$(CH_3)_3C-\ddot{O}-\overset{\overset{:\ddot{O}}{\|}}{C}-\ddot{N}H-\underset{CH_3}{CH}-COO^- \quad + \quad {}^-:\ddot{O}-\overset{\overset{:\ddot{O}}{\|}}{C}-\ddot{O}-C(CH_3)_3$$

4.

$$(CH_3)_3C-O-\overset{\overset{O}{\|}}{C}-NH-CH_2-\overset{\overset{:\ddot{O}}{\|}}{C}-\ddot{O}:$$

$$C_6H_{11}-\ddot{N}=C-\ddot{N}^--C_6H_{11}$$

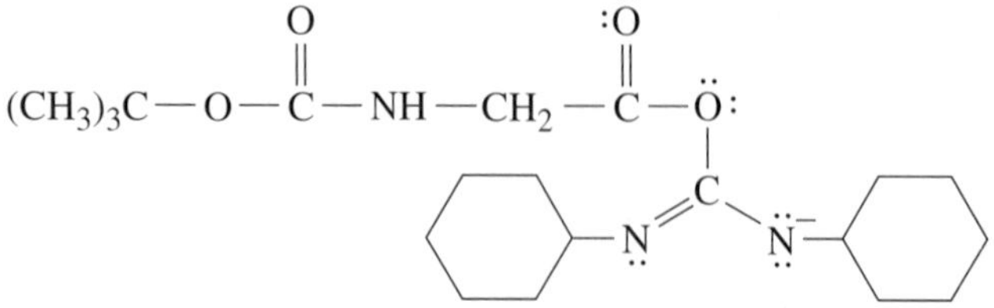

5.

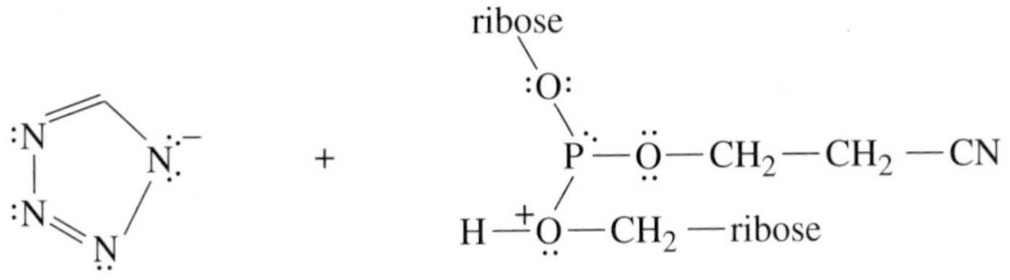

:N N:⁻ :N N

+

ribose
:O:
P — O — $CH_2$ — $CH_2$ — CN
H —$^+$O — $CH_2$ — ribose

6.

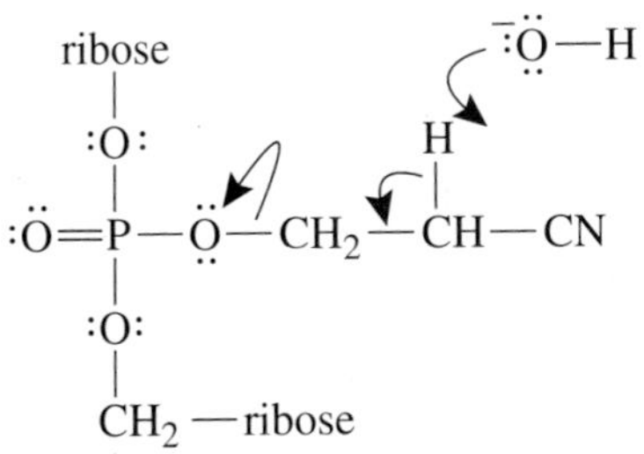

ribose
:O:
:O=P — O — $CH_2$ — CH — CN
H
⁻:O — H
:O:
$CH_2$ — ribose

7. (a)

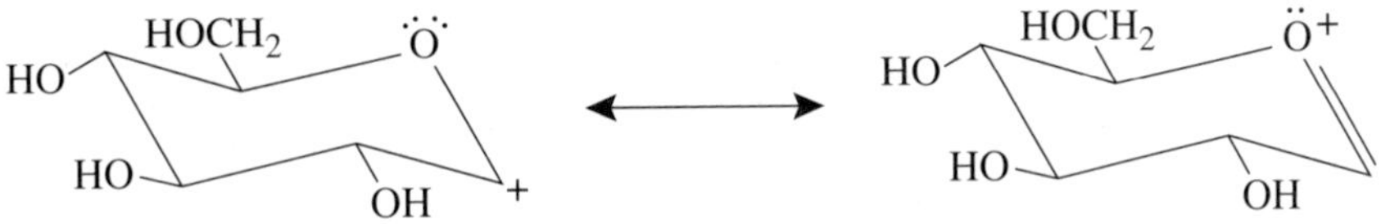

$HOCH_2$ O: HO HO OH +

$HOCH_2$ O+ HO HO OH

(b) HO $HOCH_2$ O+ HO OH

H O $CH_2$ — $CH_3$

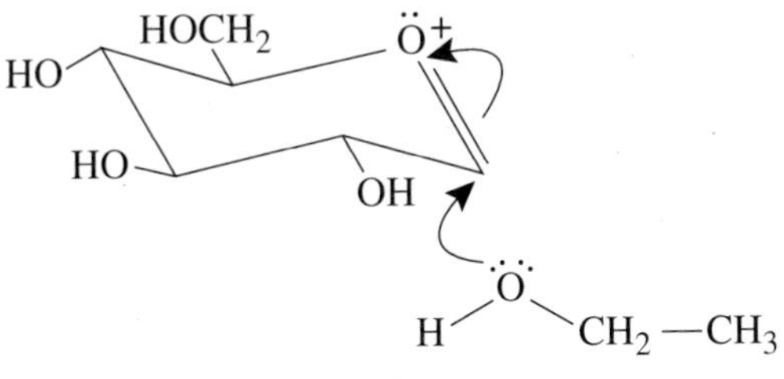

(c)

H H +O: $CH_2$ — $CH_3$

+

$HOCH_2$ O: HO HO OH :O: $CH_2$ — $CH_3$

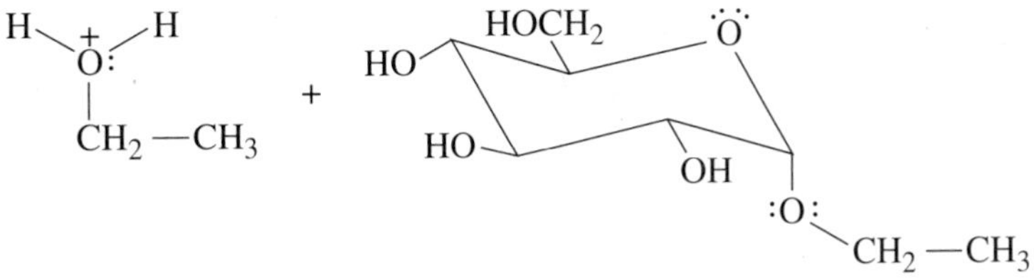

8.

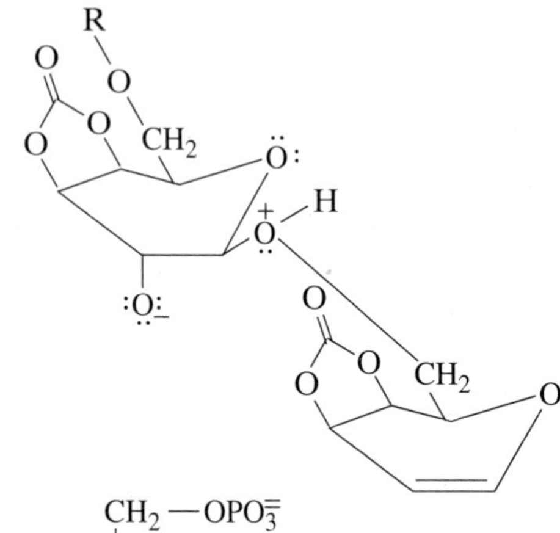


9.

$CH_2-OPO_3^=$
$C=\ddot{O}:$
$H\ddot{O}-CH$
$HC-\ddot{O}H$
$HC-\ddot{O}H$
$CH_2-OPO_3^=$

10.

$^-:\ddot{O}-P(=\ddot{O}:)-\ddot{O}:^-$
$:\ddot{O}:$
$CH_2$
HO, O, HO, OH, OH

11. (a)

$\overset{+}{B}-H$ $\quad \ddot{O}=C-O-PO_3^=$ $\quad :\ddot{O}H_2$ $\quad :B$
CHOH
$CH_2-O-PO_3^=$

(b)

$\ddot{O}=C-\ddot{O}H$
CHOH
$CH_2-O-PO_3^=$

12.

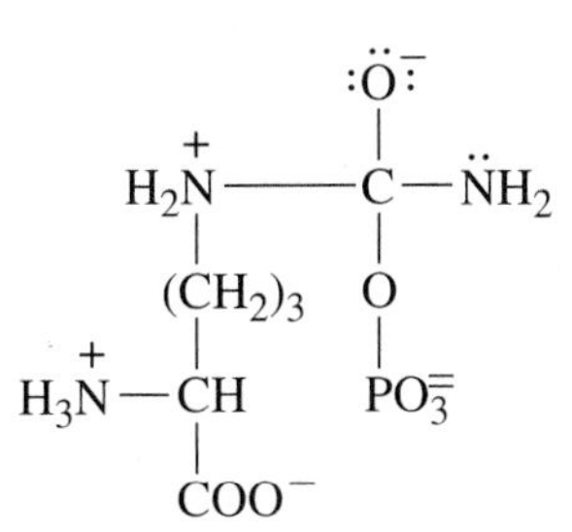

$$O^{-}$$
$$H_2\overset{+}{N} - C(NH_2)(O^{-}) - O - PO_3^{=}$$
$$(CH_2)_3$$
$$H_3\overset{+}{N} - CH$$
$$COO^{-}$$

13.

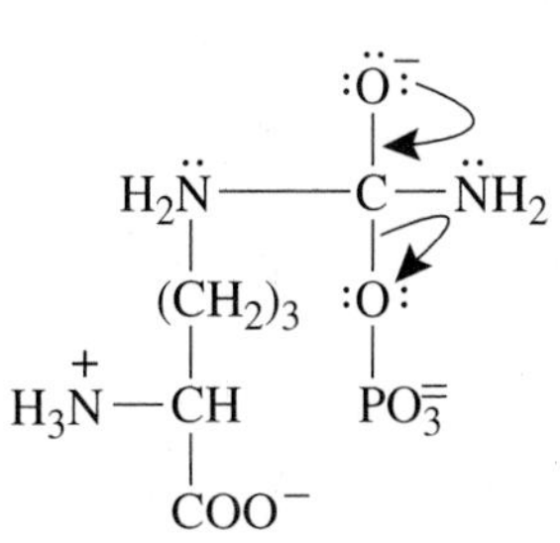

$$O^{-}$$
$$H_2N - C(NH_2)(O^{-}) - O - PO_3^{=}$$
$$(CH_2)_3$$
$$H_3\overset{+}{N} - CH$$
$$COO^{-}$$

14.

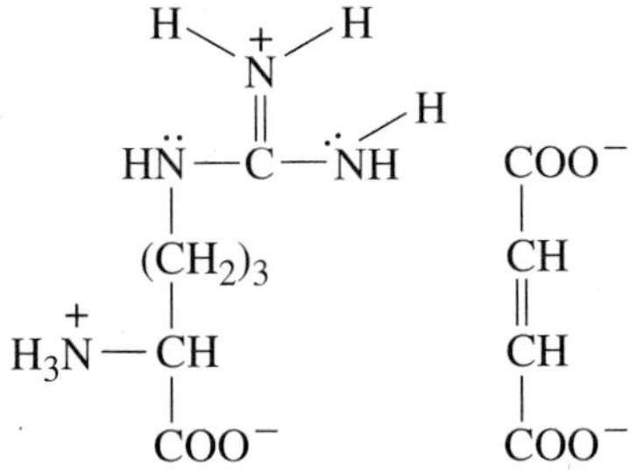

$$H_2\overset{+}{N}$$
$$HN - C - NH - H$$
$$(CH_2)_3$$
$$H_3\overset{+}{N} - CH$$
$$COO^{-}$$

$$COO^{-} - CH = CH - COO^{-}$$

15. (a)

$$CH_2 - O - C(=O) - R$$
$$H - O - H$$
N, CH, HC, C, N–H, $CH_2$
O=C–$O^{-}$

(b)

$$CH_2 - O - C(O^{-})(O - H) - R$$
H–N, CH, HC, C, N, $CH_2$
O=C–O–H

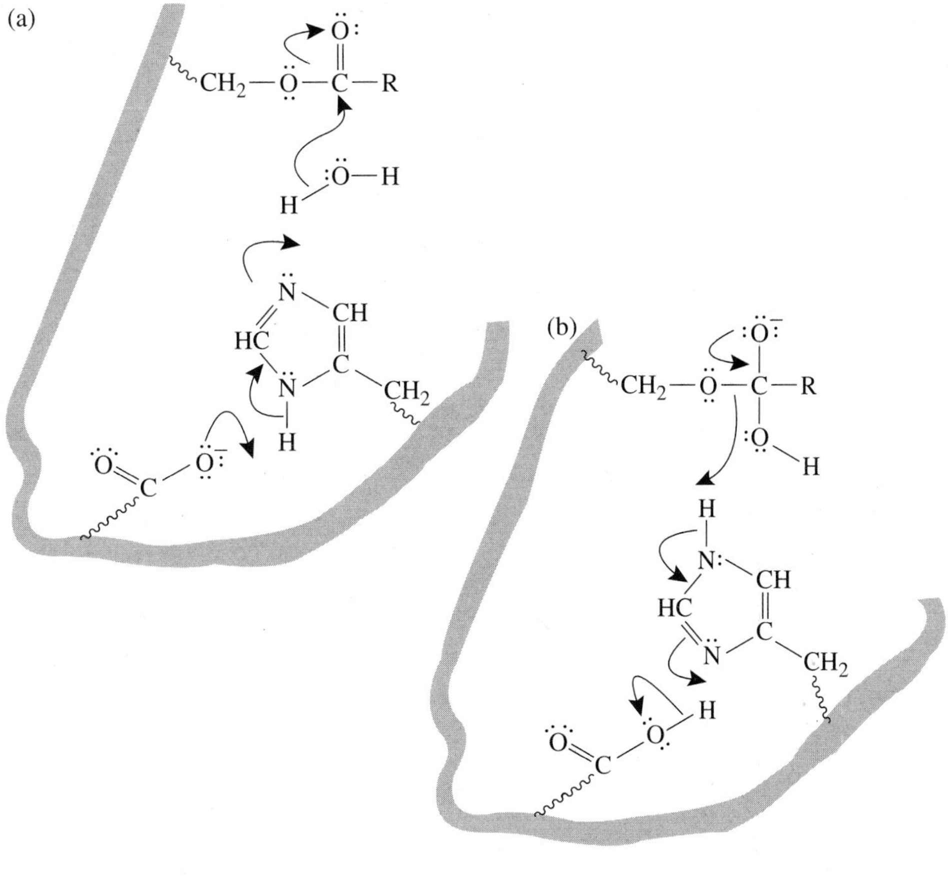

16. (a)

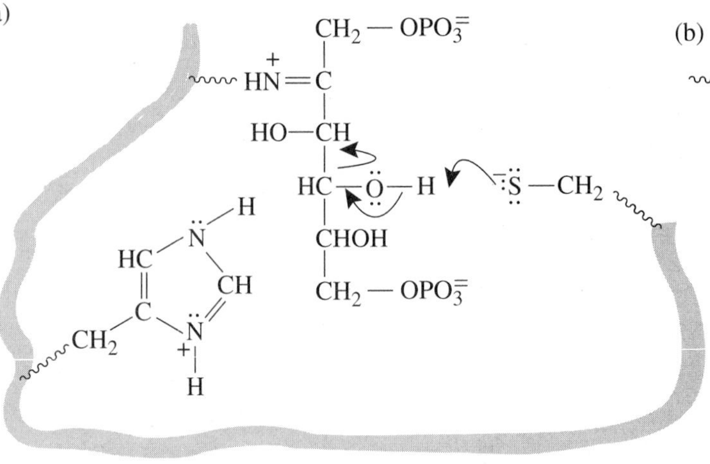

$CH_2-OPO_3^=$

$\sim\sim HN^+=C$

HO—CH

HC—Ö—H  $^-$:S̈—$CH_2$ ∼∼

CHOH

$CH_2-OPO_3^=$

H

HC N

CH

C N

$CH_2$ + H

H

(b)

$CH_2-OPO_3^=$

∼∼ HN̈=C

HO—CH

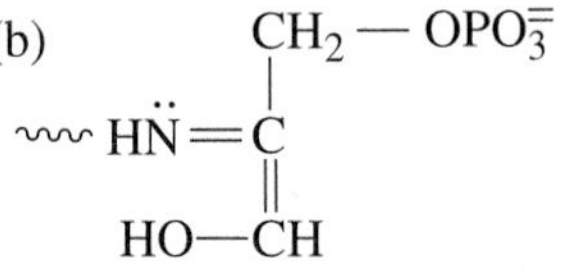

resonance structure

17.

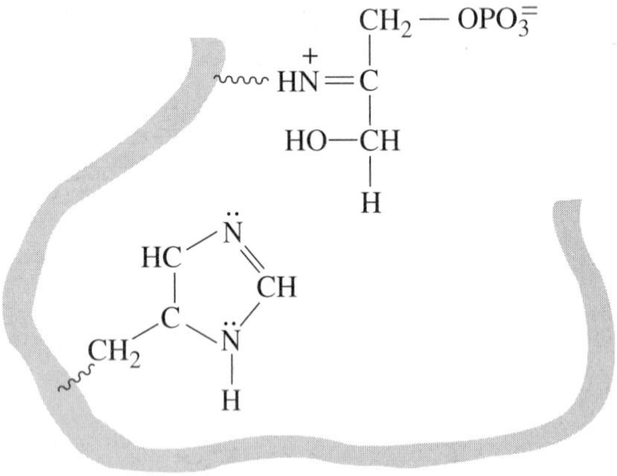

$CH_2-OPO_3^=$

$\sim\sim HN^+=C$

HO—CH

H

N̈

HC CH

C N̈

$CH_2$ H

H

18.

$CH_2-OPO_3^=$

CHOH

∼∼S̈—C—Ö:

H H

B̈ ∼∼

$\sim\sim \overset{+}{B}-H$

$CONH_2$

$N^+$

R

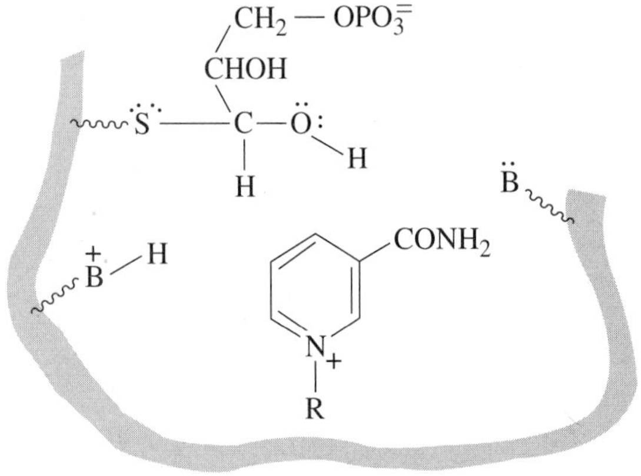

19.

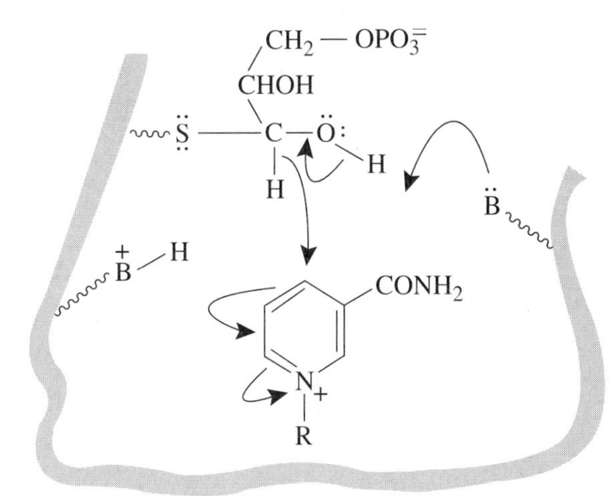
CH2 — OPO3=
CHOH
S
C
O
H
H
B
B
H
CONH2
N+
R

20.

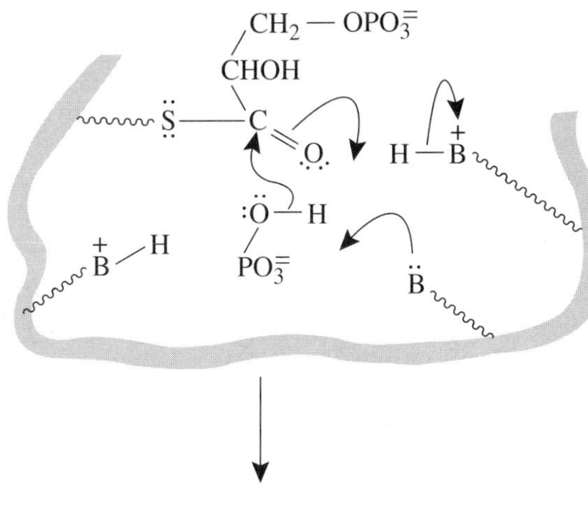
CH2 — OPO3=
CHOH
S
C
O
H — B
O — H
B — H
PO3=
B

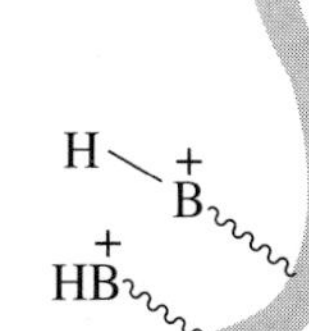
H
B
HB

21.

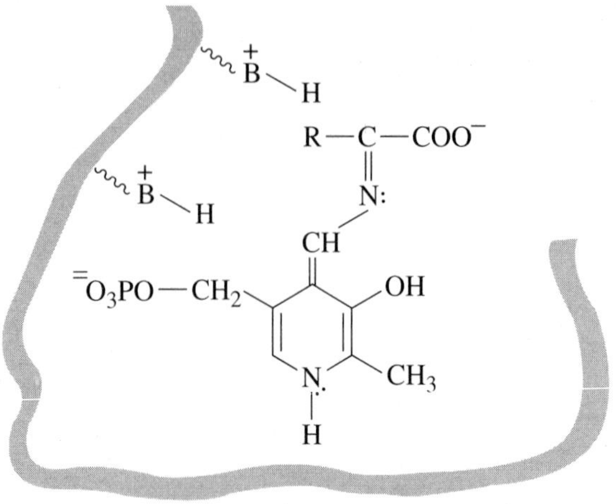

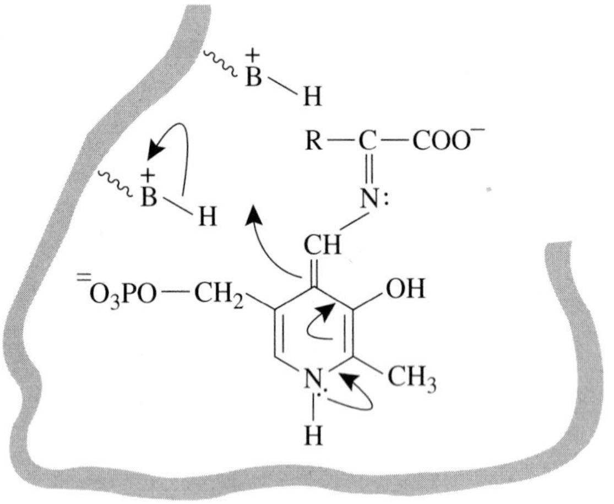

22.

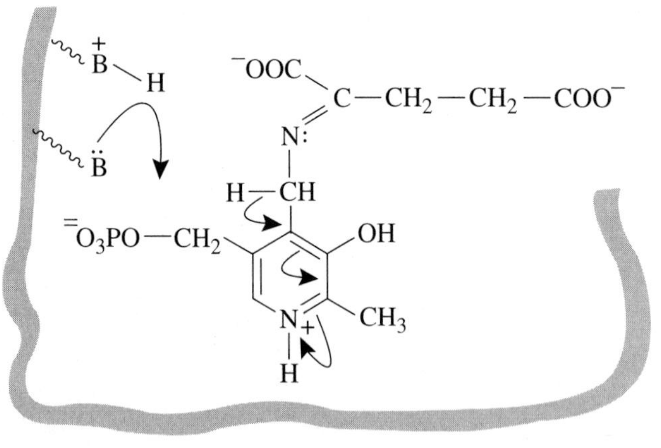

23.

23. *Continued*

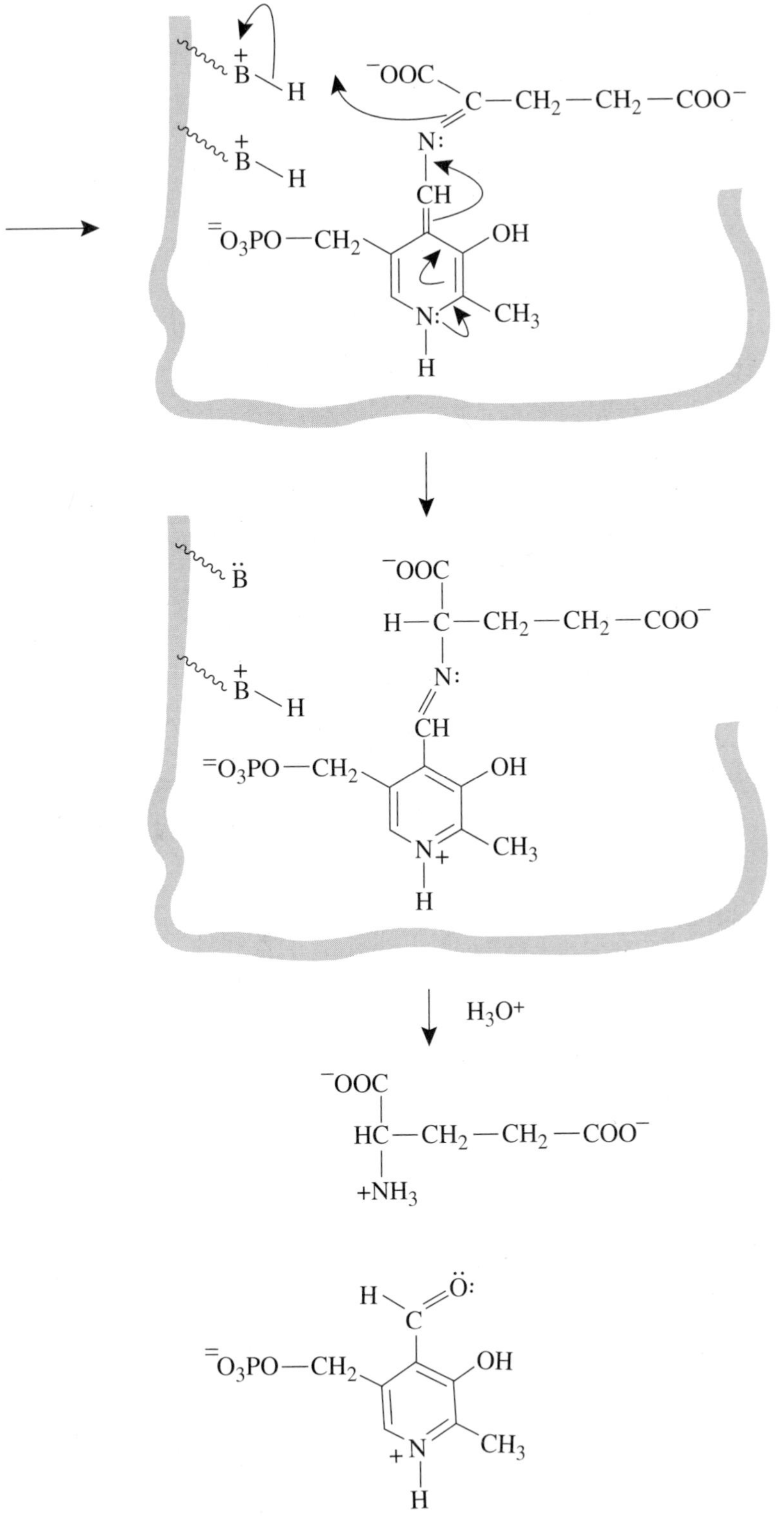

$\xrightarrow{H_3O^+}$